# SPEAKING OF STANDARDS

# SPEAKING OF STANDARDS

Rowen Glie, *Editor*
Chairman of the Education Committee
Metropolitan New York Section
Standards Engineers Society

Sponsored by the Metropolitan New York Section
of the Standards Engineers Society

**Cahners Books,** *Division of Cahners Publishing Company, Inc.*
89 Franklin Street, Boston, Massachusetts, U.S.A. 02110

International Standard Book Number: 0-8436-0307-0
Library of Congress Catalog Card Number: 70-185561

Printed in the United States of America.
The Maple Press Company, York, Pennsylvania, U.S.A.

# CONTENTS

# BIOGRAPHICAL SKETCHES

Note: Between the time we received from the authors permission to use their papers in our book *Speaking of Standards* and the actual printing of the book, things happened in the engineering profession: a tremendous turnover in jobs and positions and many relocations. It became difficult to locate the authors and to ask them for up-to-date biographies.

The necessity to come out as soon as possible with an authoritative book on standardization made it impractical to delay the publication of the book. Therefore, the biographies, in most cases, are the ones we had when the papers were presented or the article written. We hope in the second edition to correct the situation.

Rowen Glie

(Numbers in parentheses following the authors' names refer to the pages on which their articles appear.)

**W. F. ARNOLD** **(194)**

Staff Assistant, Machine Technology, IBM General Products Division Development Laboratory, Rochester, Minnesota.

**S. B. ASH** **(267)**

S. B. Ash, assistant to director of technical services, Sperry Gyroscope Division, Sperry Rand Corporation, has devoted most of his career to standardization work and materials and components engineering. He holds three degrees in engineering. He is the author of numerous articles dealing with this professional specialty and has lectured at New York University. He has been active in a number of professional and technical societies.

**CHARLES J. EIWEN** **(75)**

He is a fellow of the Standards Engineers Society and former editor of *Standards Engineering*. Mr. Eiwen was a standards engineer with the U. S. Department of the Navy and the U. S. Department of Defense, and then a standards engineer and program manager for the Alexandria Division of the American Machine & Foundry Company. Currently he is president of Daubers, Inc., a company engaged in electrical supply and servicing activities.

## RALPH E. FLANDERS (221)

Senator Flanders was born in Barnet, Vermont, September 28, 1880. He married Helen Edith Hartness on November 1, 1911. He graduated from Central Falls (R.I.) high school and International Correspondence School (Engineering) and holds 18 honorary degrees from colleges and universities.

He was a machinist apprentice and draftsman at Brown & Sharpe in Providence and Woonsocket, R. I., 1897-1901. In 1903 he was a designer at International Paper Box Machinery Co., Nashua, N. H. After serving as associate editor for *Machinery*, 1905-10, he became engineer for Fellows Gear Shaper Company, Springfield, Vt., in 1910. He was director and manager, Jones & Lamson Machine Company in Springfield, Mass., 1921, and president, 1933 through 1946.

He was United States senator from Vermont 1946-1958.

He served as president of the following: Federal Reserve Bank of Boston, 1944-46; American Research & Development Corporation, 1946; American Society of Mechanical Engineers, 1934; and National Machine Tool Builders Association, 1923.

Senator Flanders was a member of the National Screw Thread Council, 1920-24, and chairman of Screw Thread Standardization Board of American Standards Association until 1944. He is an Honorary Life Fellow of Standards Engineers Society and author of 8 books.

## GEORGE T. FRENCH (226)

George T. French, senior vice president, Overseas Operations, has direct operating responsibility for all John Deere marketing and manufacturing activities outside the United States and Canada. A director of Deere & Company since 1951, Mr. French has been with the Deere organization for 34 years.

Prior to assuming responsibility for all overseas operations in 1966, Mr. French had been in charge of worldwide manufacturing. He has held a succession of positions in manufacturing since beginning with the John Deere Plow Works in Moline, Illinois, in 1934.

Mr. French was born in Davenport, Iowa, on October 6, 1911. He was educated at Phillips Andover Academy and Yale University, graduating with an A.B. degree in 1933. He is married and resides in Rock Island, Illinois. He is a director of the U.S.A. Standards Institute, member of the American Society of Agricultural Engineers, a director of Central Engineering Company of Davenport, Iowa, and a trustee of funds and property of the Episcopal Church, Diocese of Quincy.

**JOHN GAILLARD** (15)

John Gaillard was born in The Netherlands, and received the degrees of mechanical engineer and doctor of technical science from the Institute of Technology, in Delft. For two years he was the associate director of The Netherlands Standards Association. In 1923 he came to the United States and for about a year was associated with Frank B. Gilbreth as a consultant on industrial management. In 1924 he became the mechanical engineer on the staff of the American Standards Association, and later the secretary of the ASA Mechanical Standards Committee.

Since 1947, Dr. Gaillard has been a lecturer at Columbia University, giving courses on standardization in the Department of Industrial Engineering. Twice a year, he holds a private seminar for representatives of industrial concerns on the organization of company standardization work and the art of writing standards specifications.

Dr. Gaillard is a member of the American Society of Mechanical Engineers; the American Society for Quality Control; the Society for Advancement of Management; and the Standards Engineers Society.

Dr. Gaillard is now retired and resides in The Netherlands.

**ROWEN GLIE** (22, 91)

Rowen Glie, mechanical engineer, was design engineer with Maxson Electronics Co., then the manager of the Maxson Standards Department. Later, he became the manager of the Engineering Standards Department of Kollsman Engineering Corporation. Presently, Rowen Glie is the consultant engineer with the Anaconda Wire and Cable Company. He has written many papers on standardization, engineering administration, and cost reduction, which appeared in the national technical and engineering press, such as: *Machine Design, Electrical Design News, Graphic Science, Magazine of Standards* of ANSI, *Standards Engineering* of SES, etc. He has presented many papers at the annual meetings of the Standards Engineers Society.

He served as a sectional director and as a chairman of the New York Metropolitan Section of SES, was twice chairman of the SES national meetings. Presently, he is the chairman of the Educational Committee of New York Metropolitan Section and a fellow of the Standards Engineers Society.

**MADHU S. GOKHALE** (15)

Madhu S. Gokhale is standards coordinator in the Central Engineering Department of the Radio Corporation of America and has been active in standardization work within the company and in various industry and military committees since 1939. He has participated in educational activities in standardization at RCA, Temple University, University of Pennsylvania, and in seminars at other institutes. During 1962–1963, he was appointed a technical expert to the Government of India by the United Nations, upon leave of absence from RCA. After completion of this assignment, he coauthored, with Dr. John Gaillard, a U.N. document on industrial standardization, which has been used as a text. Mr. Gokhale is a charter member of the Standards Engineers Society; was its vice-president, 1953–1954, president, 1955–1956, and was elected a fellow in 1958.

**V. de P. GOUBEAU** **(201)**

Mr. Goubeau is deceased and no biography is available.

**JOHN HERBERT HOLLOMON** **(3)**

Mr. Hollomon was graduated from Augusta Military Academy, Fort Defiance, Virginia. He received a B.S. degree from Massachusetts Institute of Technology in 1940 and a Ph.D. in 1946. He was also awarded an honorary Ph.D. in engineering by Worcester Polytechnic Institute.

He was associated with General Electric Research Laboratory-Metallurgy & Ceramics Research; then general manager, General Engineering Laboratory; assistant secretary for science and technology, Department of Commerce; adj. professor, Rensselaer Polytechnic Institute; consultant to President's Science Advisory Committee.

Mr. Holloman served as major in the U.S. Army, 1942 (Chief Physicist, Metallurgy Section, Watertown, Mass. Arsenal). He was decorated with the Legion of Merit; received the Rossiter W. Raymond Award of American Institute of Mining and Metallurgical Engineers, the Alfred Noble Award (combined engineering societies) 1947, and the Rosenhain Medal, British Institute of Metals. He is a fellow of the American Academy of Arts and Sciences and a member of many other scientific societies.

A founder of National Academy of Engineering, he is an author of many works on metallurgical subjects and an editor of and contributor to many scientific publications.

He is presently president of the University of Oklahoma.

**GERALD C. JOYNT** **(182)**

Mr. Joynt entered General Motors Technical School in 1928; he graduated in 1932. He was a supervisor in the production engineering department of Oldsmobile Division, General Motors, 1932 to 1946. He was engaged in the building business in Arizona, 1946–1952, then joined Ford in 1952 as manager manufacturing engineer of Highland Park Division-Ford Motor Company. Manager of Trenton Division 1959 – Ford Regional Manufacturing, Manager of International Manufacturing (Africa, Australia, New Zealand, and Malaysia) – Ford. He has been manager of machine process-manufacturing staff, 1964 to present, at Ford.

**EDGAR S. KIRK** **(156)**

Edgar S. Kirk, a native of New York City, is presently a standards engineer with Bostitch, Division of Textron, East Greenwich, Rhode Island. In this capacity he is responsible for materials specifications, standards, technical cataloging, and engineering procedures.

**FULTON R. MAGILL** **(105)**

Mr. Magill is a fellow of Standards Engineers Society. He retired as manager of Central Standards, Rockwell Manufacturing Company, and is presently on special assignment at Mine Safety Appliances Research.

He has had articles published in technical, educational, and trade association magazines; has lectured at Mellon Institute of Research, Carnegie Institute of Technology, and University of Pittsburgh.

He was educated at Carnegie Institute of Technology in design, with graduate work in management, and had courses in industry at University of Pittsburgh. He holds membership in Society of Automotive Engineers; American Society of Testing and Materials; Standards Engineers Society; and Sigma Phi Epsilon and Alpha Rho Chi fraternities.

**JOHN T. MILEK** **(129)**

Mr. Milek is a member of the technical staff of Hughes Aircraft Company, where is is associated with the Air Force Electronic Materials/Properties Information Center. He is the author of a reference text, *Guide to Foreign Sources of Metallurgical Literature*, and more than 60 publications in major engineering journals, two of which have won awards. He also teaches courses in technical specification writing at Santa Monica City College.

A member of Society of Technical Writers and Publishers Inc. since 1963, he has contributed to STWP publications, appeared in chapter programs, and served as chairman of the Education Committee. He holds a B.S. degree in chemistry from the College of the City of New York, with postgraduate courses at a number of universities. He is a fellow of the Standards Engineers Society and a member of the American Society for Metals, the Institute of Electrical and Electronic Engineers, and the Society of Aerospace Materials and Processes Engineers.

**IRVING O. MINER** **(111)**

No biography available.

**J. V. PANEK** **(75)**

No biography available.

**PAUL A. PIPER** **(286)**

Paul Piper is director of engineering standards, responsible for directing and administering the interdivisional standardization programs of the Aerospace Group of Martin Marietta Corporation. During more than 30 years at Martin Marietta and the Martin Company his career has spanned a variety of technical positions including: Titan II Program Planning, P6M Design Support, Project Engineer for Commercial Aircraft, 202 and 404, and earlier experience in structural design and weight engineering.

He received a B.S. degree in electrical engineering from Michigan State University and has done graduate study at the University of Michigan and Drexel Institute. He is a member of Tau Beta Pi, Standards Engineers Society, and the Institute of Aeronautics and Astronautics.

**LESLIE DANIEL PRICE** **(169)**

Mr. Leslie D. Price was born in 1901 in Manchester, New Hampshire. He has a bachelor of science degree in general engineering and is a registered professional engineer in Maine and New Jersey. He started working as an assistant electrical engineer for the Underwriters Laboratories, Inc. He later joined the Public Service Electric and Gas Company as a meter and wiring engineer. His last position before retiring was with the National Electrical Manufacturers as manager of engineering and the Safety Regulations Department. He retired in 1966 and became a consultant in the special field of national and international industrial and safety standardization (mostly electrical). In 1965 he was the recipient of the Standards Engineers Society's Leo B. Moore Medal.

**JOHN L. RAYFIELD** **(97, 161)**

Mr. John L. Rayfield, a native of New York, is presently manager of product quality for the EDO Corporation and has eight different sections under his direction: standards, (including standards library), production liason, provisioning technical documentation, configuration, mechanical inspection, electrical inspection, and quality analysis.

Mr. Rayfield joined the EDO Corporation in 1963 as standards section head, and was assigned to his present position and duties in 1967. He formally held positions in other firms as standards section head, manufacturing supervisor, executive engineer, and in many other capacities. He is sectional director of the New York Metropolitan Section of the Standards Engineers Society; a member of ASQC, AIDD, IEEE, and member of the New York City Council on Education and Programs.

**RICHARD L. M. RICE** **(208)**

Richard L. M. Rice, standards engineer in the design division of the Du Pont Company's Engineering Department, Wilmington, Delaware, has been engaged in engineering work for the company since 1950.

Mr. Rice joined the Du Pont Company as a management engineering consultant. He served as a process engineer in the design of the Savannah River Plant for the Atomic Energy Commission and in 1953 was assigned to the standards section. In this position he has contributed materially to the broadened use of engineering standards throughout the company.

Born in Philadelphia, Pa., December 31, 1920, Mr. Rice attended Cornell University where he received his B.S. degree in administrative mechanical engineering in 1943. He served two years in the U.S. Army and spent five years as an industrial engineer with Chance Bought Aircraft Company before joining Du Pont.

Mr. Rice is a member of the Standards Engineers Society, currently serving as national director from the Philadelphia Section. He is a member of the National Society of Professional Engineers and is a registered professional engineer in the state of Delaware.

**GEORGE W. RITTER** **(230)**

Mr. George W. Ritter, a native of Bluefield, Virginia, is presently assistant commissioner, Standards and Quality Control, Federal Supply Service, General Services Administration. In this capacity he is responsible for the specifications, standards, cataloging, and quality control programs of GSA, as well as the direction of a material evaluation laboratory. He was formerly an official of the Department of Defense, where he served in various high level positions. He served as an officer in the U.S. Navy from 1942 to 1950 and presently holds the rank of Captain, USNR (Retired).

**ARNOLD M. ROSENWALD** **(145, 216)**

Mr. Rosenwald is supervisor of the standards section, Philco-Ford Corporation WDL Division, Palo Alto, California. He is responsible for the

WDL standardization program in design, parts and materials, manufacturing standards, and drafting and documentation.

He has had extensive experience in the field of industrial standardization as section manager with the Raytheon Company and Tracerlab, Inc. He has been a regular columnist and contributor to the magazine *Standards Engineering*, a publication of the Standards Engineers Society, and has received citations for contributions to both the magazine and to SES.

Mr. Rosenwald studied at the Lowell Institute, (an affiliate of MIT) in the mechanical, electrical, and business courses. In addition to his basic education, he has studied in the fields of psychology, economics, business English, and production and participated in standardization seminars conducted by Dr. John Gaillard and Professor Leo B. Moore.

He has addressed many professional groups and seminars on the techniques and advantages of standardization and has served in all elective offices of the Boston Section Standards Engineers Society and on the SES Board of Directors. He is presently chairman of the San Francisco Section and future planning committee chairman of the society.

**JOEL G. WALKER** **(256)**

Joel G. Walker is a native of New York City but has lived for many years in metropolitan New Jersey. He has more than 40 years of service with Bell Telephone Laboratories, where he is a member of the Engineering Information Center at the company's Whippany laboratory.

Early in his Bell Laboratories career Mr. Walker was concerned with the installation, operation, and maintainance of reproduction equipment. He later worked on the installation and operation of a magnetic materials heat treating laboratory.

During World War II he worked in conjunction with the National Defense Research Commission and the Naval Ordnance Laboratory on the development of magnetic devices for the Navy Department. He later assisted in studies involving metallurgical X-ray and metallic crystal technology. During this period he coauthored several articles on metallurgical crystallography.

Mr. Walker's background also includes some time with the personnel department, doing job analysis and evaluation. His present responsibilities include documentation standards used in association with government projects.

Mr. Walker is a member of the National Microfilm Association, the New York Section of the Standards Engineers Society, and serves on subcommittees of the United States of America Standards Institute (USASI) and Bell System standards committees. He has presented papers to technical and standards societies on microreproduction, standards libraries, microrecording processes, and establishment of record systems.

**EVERETT WOERTER** **(79, 120)**

Everett Woerter is vice-president of Jay Bergen & Associates, Inc., a consulting firm that was founded in 1960. Mr. Woerter has extensive experience in the field of engineering practices and procedures. His experience includes more than 20 years in drafting, design, standards, and engineering administration.

Mr. Woerter has presented numerous papers on subjects related to standardization. Many of the papers were published in the trade journals.

Mr. Woerter was associated with the International Business Machine Corporation as manager of components division standards and specifications.

He served at staff engineer for the parts division of the Electronic Industries Association, and also was associated with American Machine and Foundry Company in management and supervisory positions, including corporate manager of standards and design group leader.

Mr. Woerter has been the national treasurer of the Standards Engineers Society (1960-1962) and national president of SES (1966-1967). He is presently a member of the United States of America Standards Institute CMC Standing Committee on Commercial and Industrial Standardization.

# THE PRESIDENT'S OBSERVATIONS

Never, in the long evolution of standardization principles and the effective application of the rationalization concept, has this tool, so beneficial to all mankind, received such extensive publicity, both favorable and otherwise, as at the present time.

As a result of this attention, many opinions have been expressed through articles, government hearings and briefs, industry presentations, seminars, etc., by informed associations and individuals and also, unfortunately, by many who are not so knowledgeable of the "standardizations business."

The timeliness, therefore, of this reference source by qualified authors covering various technical and administrative aspects of standardization is particularly appropriate.

Not only do I hope that the papers themselves will provide some answers to present-day standardization problems, but that they will also provide points of contact for additional information.

For it is through American National Standards Institute and other national standardization associations, including the Standards Engineers Society, that qualified and expert sources of knowledge can be found, when required, on the very important subject of standardization and its effect on our economy, both present and future.

Michael J. McKerrow
President 1969-1970
Standards Engineers Society

# FOREWORD

In an exploding technology such as ours, the need is rapidly growing for practical expert guidance and reference information for improving standardization in all fields. The objective in making this book available is that some useful portion of this need may be met by this first compendium of skillfully selected and edited papers on several basic aspects of standardization.

The papers were chosen and arranged to form a coherent composite, but any one may be treated independently as an individual entity of standardization if the reader so prefers. While recognizing and implementing the standardization concepts which are contained herein, the reader would be well advised to keep in mind a fundamental principle of standardization: to provide in every standardization program, document, or procedure a sufficient flexibility to permit future improvements as new ideas, inventions, or states of the art become available.

The authors who have made this volume possible represent an unprecedented massing of real experience, mature judgment, and responsible positions in widely varying aspects of standardization, including industry, government, and education – international in scope. These authors are not only unusually articulate, but they have provided a unique quality of first-person directness arising from the personal style necessary in oral presentations to small but critical audiences.

The editor, Mr. Rowen Glie, of Anaconda Wire and Cable Co., brings the wisdom of a long professional career in standardization to this present work, reflecting his several decades of directing, lecturing, writing, and consulting in standardization affairs. He and his associates in the New York Section of the Standards Engineers Society are to be commended for their effort and dedication in preparing this volume.

While most of these selections are engineering oriented, the editor has done a service in "making this book available to all those interested in standardization," not merely to standards engineers. Persons of every discipline, including but not limited to technology, engineering, and science, in governmental, industrial, commercial, consumer, and other areas, may find here some insight for doing a better job at less cost.

Jackson K. Emery
President 1968–1969
Standards Engineers Society

# FROM THE EDITOR

The often repeated assertion that information about standardization is not available is not a truism but a paradox. There is a great amount of information about standardization which is available. However, it does not come to us in the form of books, but in the form of articles and papers presented at meetings, conferences, symposiums of different engineering and industrial organizations or it is published in technical magazines.

In these papers and articles standardization is considered from all points of view – national, international, military, industrial, technical, that of standards engineers, of management, and that of users. The most important sources of data on standardization are the articles which appear in *The Magazine of Standards*, published by The American National Standards Institute (ANSI), (formerly the USA Standards Institute); in *Standards Engineering*, the magazine of Standards Engineers Society (SES), and in papers presented at the annual, semiannual, and monthly meetings organized by these two societies, as well as publications issued by different engineering, trade, and purchasing agent organizations and the Department of Defense and Commerce.

By the decision of the Board of Governors of the Metropolitan New York Section of Standards Engineers Society, I was given the task of gathering into a few volumes selected material on the subject of standardization that will present a comprehensive picture of the purposes, results, and practices of standardization. By making this book available to all those interested in standardization and selling it to society members at cost, we hope to achieve a better dissemination of knowledge and understanding for standardization, not as a process destined to stifle imagination and creativity of engineers, but as a process natural to a highly developed stage of technology. Standardization should be a means to accomplish maximum effectiveness in creativity, honesty, and understanding between engineers and shops, vendors and buyers, manufacturers and customers, importers and exporters; a means of insuring a high standard of living; and a method of making national defense more effective by reducing its logistical problems. It should be a system for assuring to everyone a participation in technological progress by making engineering management, design, and R & D decisions based on facts and enable thereby reliable managerial predictions. Finally, standardization is a major factor in promoting national and international trade.

There are few engineering specializations which serve so many noble causes, all at once, as the standards engineers profession.

Because standardization, among other things, is somehow very much overlooked in engineering education it is not surprising how little is known and even less understood of what standardization truly means. It is not realized that standardization is always present when people are organized to achieve a common goal, when know how must be used, when things must be predicted and people have to talk to and comprehend one another.

But, there is a great difference between a rationally organized approach to standardization and an intuitive one.

Birds can fly, build nests, and navigate without learning; trout and giant turtles are also great navigators without the benefits of the Naval Academy at Annapolis; beavers build dams without studying civil engineering or even passing a professional engineers examination. They do all this by instinct. In the same way – by instinct – too many standardizers work in thousands of companies, in the Department of Defense, in industry, in national, technical, trade, and engineering societies.

It may be strange that such a fundamental technical function as standardization is still done by instinct and trial and error method, in darkness as to its purposes and methods. It is strange, but it is a fact.

This book tries to throw light on the four facets of standardization: (1) learning or gathering of knowledge, (2) selecting or organizing knowledge, (3) promoting the use of standards or utilizing knowledge, and (4) communicating or spreading knowledge and know how.

We sincerely hope that engineering students interested in standardization as their career, standards engineers looking for a solution to some of their problems, and engineering management which seeks clarification of what standardization is and what can one expect from it will find in these pages the answers they seek.

The first volume of this series is a collection of articles presenting the ideas and general know-how of the most authoritative specialists in standardization. To our knowledge there is no book on standardization which presents between its covers an equal amount of professional competence.

The articles selected for inclusion here have originated over a period of several years, but they all have one feature in common: timelessness. What was true at the time they were written is still true at the present time. The problems in standardization and its solutions are still the same.

In order to achieve a sense of unity in this book, only the names of the authors are provided here. The specific dates and sources of the articles are purposely omitted to emphasize their timelessness and timeliness. Should such information be needed, the editor would be glad to supply it.

However, everything that is published in this first volume is taken from the *Magazine of Standards, Standards Engineering*, ANSI National Company Conference publications, the material published by the Standards Engineers Society as given at their National and Sectional Meetings, and from the UN publication *Industrial Standardization in Developing Countries*.

It is obvious that this book would be impossible without the help which this editor received from many people, to whom I would like to express my personal gratitude: Mr. J. Boucher, Mr. and Mrs. C. Euffa, Mr. N. Gutman, Mr. W. V. Weyer, Mr. H. Reiss, Mr. J. Rayfield, Mr. J. Boese, Mr. A. E. Smith, Mr. J. Walker, and Mr. T. A. Herman of the Board of Governors of the Metropolitan New York Section, who provided financial and, what is more important, moral help and encouragement to the editor.

For permission to reprint articles which appeared in their publications, our thanks go to: Mr. Donald L. Peyton, the managing director and secretary of the American National Standards Institute. Our thanks also go to the members of ANSI Staff, Miss Dorothy Hogan and Mr. B. Stanleight, for the help they extended to us and for permission to use material published by ANSI; to Cahners Books for publishing and promoting a book on standardization and, of course, to all the authors and speakers who by writing their articles and presenting their papers shared their professional experiences and knowledge with all who wanted to learn. Also I would like to express particular appreciation to the Executive Committee of the Standards Engineers Society in the person of their presidents, Messrs Jackson K. Emery, the past, and Michael J. McKerrow, the present, for sponsoring this publication.

Rowen Glie

# Part 1

# INTRODUCTION AND DEFINITIONS

*The old sage, Socrates, liked to start any discussion with the request: "Define your terms." This is a sound suggestion – unless we know what we are talking about, what are we truly talking about?*

*In this section we will try to define standardization and standards. This is not as easy as it seems since standards are many things to many people.*

*R.G.*

# STANDARDS AND THE PUBLIC INTEREST

J. Herbert Hollomon

In a complex mass society standards serve many purposes. They provide a basis for exchange of goods and services. They permit mass production. They measure the safety of products. They are the basis for keeping unsafe foods and drugs off the market. They establish product performance in the interest of consumers. Food, clothing, shelter, and comfort products are measured for how well they perform. Labeling standards prevent deception and warn against hazards to health. Electrical standards protect the public from hazardous appliances and electrical wiring. When standards set within different countries are harmonized internationally, channels of trade and commerce readily bridge national boundaries. Domestically, standards create a healthy environment for the widest sharing of our resources and products through the market place. Internationally, standards stimulate trade between nations. Standards serve both the highly specialized private needs of industry and the more public demands of people as consumers of products. Standards, ill-conceived or improperly processed, can also have harmful effects.

Both President Johnson and the Congress have stressed the public's need for standards which reflect the public interest. I would like to discuss three ways in which standards serve the public interest and three ways in which a standards-making body can serve the public interest.

## How Standards Serve the Public Interest

Walter Lippman in *The Public Philosophy* defines the public interest as "what men would choose if they saw clearly, thought rationally, and acted disinterestedly and benevolently."

Can we see clearly and think rationally when considering standards? I believe we can. Standards serve the public interest in at least three ways. First, fundamentally and historically, standards serve the public interest by enforcing honesty and protecting the consumer with respect to the goods and services he buys.

In our country, coinage and weights and measures were established as a responsibility of government in the Articles of Confederation, ratified by the colonies in 1781. That power was transferred to Congress by the Constitution of the United States, effective 1789. More recently, as industry and the nation have grown, reasonable standards for many products have traditionally been established through voluntary cooperative action of producers, distributors, users, consumers, and government agencies. But also, numerous state and federal laws and regulations deal directly with standards for consumer products, and various government regulatory agencies are charged with adjusting the varied and sometimes conflicting interests of various groups in the total public interest.

For six years, 1921–26 inclusive, Herbert Hoover, while Secretary of Commerce, took a great interest in standards. Standards, he said, contribute to "protecting both producer and consumer" and "to make the wealth and strength of the American people."

By and large, producers, distributors, consumers, and trade associations all have shown a strong sense of responsibility. The implementation of voluntary product standards is working well in this climate of cooperation. We now have in effect about 500 voluntary product standards developed through the Commerce procedures, and about 13,000 others developed through private associations. This great system of standardization operates in the best American traditions of individual freedom and national cooperation.

There is a growing public attitude that the standards process should not be under the sole control of a limited group whose interests are affected. As early as the 1906 Food and Drug Act, Congress acted to authorize government, rather than the interests concerned, to begin regulating unsafe food and drug products. We see the same history in all segments whenever the public interest has become clearly paramount. The regulation of securities by the Securities and Exchange Act of 1933 reflected the public interests against securities manipulations and abuses in the control of stock exchanges having economic impact on the entire country. The Meat Inspection and the Poultry Products Inspection Acts reflected the public interest in protecting the public health through safe consumer products. More recently, the mood of the country has demanded protection for the consumer against abuses in the marketplace. The cigarette labeling act reflected a concern that the individual be warned of the health hazards of smoking. The National Traffic and Motor Vehicle Safety Act of 1966 declared it to be national policy to reduce the totally unacceptable death and injury rate on the highways. The Fair Packaging and Labeling Act expressed the national policy to end confusion to consumers through undue proliferation of quantities in which consumer commodities are sold. The truth-in-lending bill is for the same purpose in preventing confusion of individuals about the terms of loans. Recently, the public interest in reducing air pollution has been stated quite clearly by President Johnson when he called for, among other things, emission standards for certain industrial plants, as well as for automobiles.

Even when government sets mandatory standards, however, it is impossible to reach the established public goals without a fair procedure for using the vast competence of industry. Only then can we expect standards that actually produce public benefits in safety or health at a realistic cost to society and without harmful consequences. Standards must be feasible from the standpoints of both cost and engineering. Industry must help determine that feasibility, and it can do so only by upgrading the competence and the professional status of the standards engineer.

Secondly standards serve the public interest by stimulating innovation and competition.

In recent months I have become more convinced than ever before that sound standards policy can actually help reach the goals of the antitrust laws and favor innovation and competition – perhaps better than antitrust lawsuits can. Let me explain why. The standards which most often can be misused are those written to exclude rather than include various products or quality of products. Exclusive standards are poor standards because they deter innovation and competition. Inclusive standards are good standards when they are written in terms of a variety of levels of performance of a product. They leave the preferred level to the user. Standards written in these terms measure total performance for certain purposes. They specify that products should be graded and labeled according to the level of performance. If the public is informed of the level of performance of the product, each consumer can choose as he will. Moreover, firms have an incentive to compete for better products meeting the levels of performance without the need for changing the standard. The marketplace selects preferences among the products meeting each performance level. If a particular level of performance is unacceptable to society because it is considered unsafe – as for example in motor vehicles or flammable fabrics – then by a public process the community can act to bar that level. Otherwise, private choices are permitted.

In some countries such as Sweden, the grading and labeling of a product's performance may be done by government certification. A different approach is being developed in England where a "Teltag" system has been introduced to tell consumers about a product's performance. The test criteria under the "Teltag" system are determined by industry in cooperation with the British Standards Institution. It does not involve government in product testing or endorsement of certain brands. The idea is also being suggested as a possibility in this country. Such a system would distinguish between writing a standard which measures levels of performance and informing buyers of the quality levels for the product sold under the standard. Competition based on the quality of products would be encouraged by such a system.

Suppose government participates in a voluntary standards process and even approves the purpose of the standard and the final result. Does that have any significance on how far private pressure is tolerated under the antitrust laws? I believe it does, but not because government has blessed the standard with sanctity. Except for mandatory standards, government participation and approval have nothing to do with compliance and thus do not create any antitrust exemption. Rather, they speak to the two essentials, due process and purpose. The fact of government participation is some assurance of unbiased competence. The fact of approval is a sign that some impartial source has assessed the value of the standard. Government representation in the voluntary standards process means that government has served as an impartial representative.

There may seem to be a paradox between stimulating innovation and competition and eliminating unsafe products. Aren't there dangers when private standards presume to exclude products falling below certain levels of performance? When you look upon standards as inclusive, written in terms of various levels of performance and not in terms of materials or a particular process, then competition is preserved. Leaving the preferred level of performance to the user, you realize that in this sense, standards do serve the public interest by stimulating innovation and competition while they protect the public by showing the level below which a product is considered unsafe.

Thirdly, referring to a theme embodied in the declaration of purpose of the international standardization bill, standards serve the public interest by stimulating and facilitating domestic and international trade.

During the last decade or so, the rapid increase in goods and services flowing across state and national boundaries demands standards specially devised to reduce unnecessary barriers to trade. The public interest is increasingly being written into transactions between two contracting parties whereever those parties do business. In the exchange of goods and services internationally as well as domestically, standards reflecting the public interest should be part of the system of exchange. Traditional laws regarding commercial transactions simply have not taken full enough account of the significance of standards in this system of exchange.

I find it most interesting to study the rapid development of modern Europe partly through the merchants who sold goods and developed their own "law merchant," the commercial code of the time. The commercial law which permitted the economic rise of modern Europe invaded the structure of the feudal system and the European nation-states from outside the system. It was, as the lawyers say, extralegal.

Today, we see a similar invasion of our system of exchange based on contract. But we have not seen clearly enough the significance of voluntary standards and their impact on our commercial system. Standing somewhere between the parties to a contract for the exchange of goods and services – yes, and between nations, too – there is increasingly a standard of performance and sometimes of safety. That is so most clearly when products are complex systems engineered to meet systems needs and serve the general class of public users – jet aircraft, automobiles, the pipeline. It also is true when products are much simpler, such as washing machines, vacuum cleaners, and the whole range of daily commodities which not only our housewives but the consumers of many nations want to buy. How these standards are developed and what they do is of public concern. This concern is partly expressed by the new idea that the consumer has as much legal status as the producers in setting standards.

These standards are of public concern not only because they have an economic impact. If national barriers arise from national standards, then political problems between nations are inevitable. In the United States, barriers between the states also may arise. These may be undue burdens on interstate commerce, and they frequently arise through standards for products. Even where our Congress has not preempted the field by passing uniform standards, the federal courts may strike down (a) any state regulation that is so arbitrary and discriminatory in its operation as to violate the due process clause of the Fourteenth Amendment to the Constitution and (b) any state regulation over interstate commerce that constitutes an undue burden on interstate or foreign commerce or that discriminates unduly against out-of-state manufacturers and shippers.

These doctrines are basic constitutional concepts in the area of standards and laws regulating the exchange of goods and services. I am advised that there is a shift in emphasis of the case law from giving preference to the right to contract to supporting the right of state and federal governments to establish law that is concerned with the functional social usefulness of particular goods and services. Formerly the courts recognized rights to contract with respect to any goods or services, with less emphasis on the overriding public interest. If I sense the shift correctly, it means that we now should be looking to weights and measures laws or product standards as the common ground on which goods are exchanged in the public interest. No longer in a complex product line is freedom of contract – caveat emptor – sufficient to continue its leavening and freeing effect. The freedom to contract produced the commerce that helped stabilize and restructure modern Europe and the West. In the nineteenth century, economic freedom was as important to the merchants and guilds as technology is to us. For the serf who was tied to the land, his new-found economic freedom gave him and others independence.

Today the consumer is tired of being a serf to an imposed system. He demands his freedom. He needs standards to help him choose in the marketplace. Even more, he insists that he should participate in the process by which those standards are issued, for in the marketplace he buys the products of our technology according to established standards. Sensible weights and measures regulations always – especially in today's complex marketplace – try to bring producer and user together at a common meeting ground.

In the United States, the doctrine of undue burden on interstate commerce seems to be moving toward national standards for exchanging goods and services, for it intervenes where there are burdensome or conflicting state requirements. Since the consumer now is receiving national status as a participant in commerce, the undue burden doctrine could well be used to reflect this status. Let me be emphatic here. I am not referring to national in the sense of federal, but rather in the sense of national uniformity to reduce burdens on commerce. As we see, when the burdens become too great, the federal courts step in to protect the country's interest in reducing barriers to a genuinely national system of exchanging goods and services.

### How Standards-Making Bodies Serve the Public Interest

Regarding standards-making bodies, can we also see clearly, think rationally, act disinterestedly, and achieve the positive goals of society? I believe we can. I believe we increasingly have been doing so since, two years ago, after a thorough study, the LaQue Advisory Panel on Engineering and Commodity Standards (whose chairman was Dr. Francis L. LaQue, vice-president, International Nickel Company) described two principal problems in standardization in this country. The first, the panel said, is the need for coordination of standardization activities, and the second, the need for strengthened voluntary national standardization through an appropriate mechanism leading to what might be called "USA Standards." The panel recommended that the primary responsibility for developing standards be left in the hands of the technical societies and trade associations which are principally concerned with developing voluntary standards.

In implementing one of the panel's recommendations, the Department of Commerce issued new procedures for the development of voluntary standards for products, processes, and materials. These procedures leave the responsibility for consenting to product standards with the various interests affected – producers, distributors, and consumers. The impartial procedures allow all interests to be expressed objectively and adjusted by consent. The Department of Commerce participates in the development of a voluntary standard when it meets certain tests of public interest and national effects and when it cannot be processed according to needs or desires of the industry by a nationally recognized private standardizing body.

The international standardization bill, which we have sent to Congress again this year as a desirable step towards reducing national barriers to trade, does not go as far as the LaQue Panel recommended, namely to authorize direct financial support for the new United States of America Standards Institute. But it does propose grants to qualified standardization organizations for participating in the international standards process and for information activities.

The LaQue Panel was particularly pointed in criticizing the inadequacy of the United States participation in international standardization. It could foresee national barriers arising to deter rather than enhance the flow of goods and services internationally. We in this room today know of many instances in which the curtain has dropped on the exchange of products simply because someone was not interested in participating in developing a standard for that product. The telecommunications field is one of the more obvious.

I am proposing today that we get to the heart of the standards question and see it clearly enough to protect the public interest. Whenever standards are exclusive rather than inclusive, have been written only by manufacturers, have not taken into account the desires of users and consumers, or do not exist where there is a clear public need to save lives or to protect public health, we can expect the public and Congress to become concerned. Congress may set the purpose of a standard and require it to be issued under mandatory sanction. Some believe the government should decide, if not enforce, all standards. Others believe that if the process is well enough understood, a better balance can be maintained by responsible private citizens and organizations with government cooperation. Where the purpose is the sale of goods and services only, shouldn't that sale be based upon standards that permit honest choices to be made in the marketplace? How better can the high purpose for a particular standard be determined than to permit the user of products it measures to have a democratic voice in saying what the standard should achieve and by choosing product by level of performance measured by the standard?

Therefore, I propose that every private organization or professional society, before it begins the development of a standard, adhere to a fundamental rule: Standards-making bodies should serve the public interest by defining the purpose of a standard as well as establishing the process for developing standards.

After all, that is what the Congress does when it finds that there should be mandatory standards for health and safety.

Once there is a need for a standard, the process determines what is to be measured and prescribed. I am not talking now about mandatory standards which government prescribes by a public process enforced by an arm of legal authority. I am talking about the way voluntary standards become the norms for exchanging products within the private government. Abuse in this process can generate much concern about the protection of the consumer and the public against unsafe products, for example.

There are as many processes for writing standards as there are for running organizations. I will discuss three.

The first process for writing standards: A group of manufacturers cooperate, with no prior procedures or rules, to set a standard for manufacturing a product. The risks inherent in this process have been considered by Assistant Attorney General Donald Turner in his recent address entitled "Cooperation among Competitors." Here the antitrust problems are at their maximum since the process tends to exclude the user and the public. Even some competitors might be excluded. This exclusive process for writing standards contains the greatest potential for abuse and tends to produce exclusive standards.

The second process for writing standards operates through an organization or professional association having prescribed procedures objectively administered by professionals. The risks inherent to this process depend on the structure of the organization, its financing, who has the dominant financial control, how committees are chosen, who is permitted to participate, by what standard consensus is achieved, how the issues are considered and brought into the open, and finally whether consumers and the general-interest segments of the committee have an effective voice in proposing and drawing up a standard. One might also ask whether there is an appeal process or a review function by which an impartial professional group can appraise the work of a committee. Procedures of professional societies and organizations have a tradition in this country in setting standards. I am not satisfied that enough progress has been made to change the inner workings of these groups into a fair public process. I would like to see more openness brought to the proceedings and greater competence actively cultivated from the point of view of the greater public in participating in the proceedings. I would like to see businessmen stop giving standards work low priority and inadequate financial support. Despite these criticisms, I think private groups can, and often do, arrive at a good voluntary standard through a pre-established fair procedure, which, if not abused, will avoid antitrust problems and can serve the public interest.

The third process for writing standards utilizes the government as the arbiter or the professional group administering the procedures. Both private interests and consumer interests are expressed through that process. An example is found in the voluntary standards procedures of the Department of Commerce which were strengthened considerably recently upon the recommendation of the LaQue Report. Though the government administers the procedures and requires full participation by users, consumers, and government agencies, the government does not write the standard itself. It may decline to publish a standard if it is adverse to the public interest. But even these procedures do not require the government to determine what the public interest is. That is left to the standards review committee, which must file a report justifying the need for a standard, stating its purpose, outlining the pros and cons concerning it and how they were resolved. The government does, however, judge when there is a consensus. Our procedures are still being tested (and I might add sorely tested in the lumber standards matter which is still a splinter in our side.) On the whole, they are evolving by experience into a sound concept of due process which I commend to the private organizations for their own consideration.

There is another point I would like to make: Standards-making bodies should serve the public interest by educating the public on the purpose, process, and significance of standards, and encouraging responsible participation.

Each professional standards organization or trade organization should undertake a thorough review of its procedures and revise them to serve the public interest under the following criteria:

1. A standards-developing committee must be balanced among producers, distributors, users, consumers, testing laboratories, government agencies, and other interests reflecting a more general point of view of the community whose purpose is served by the standards. A list of the names of committee members should be made public.
2. An open process should be provided with broadest participation from the public interest groups, and productive attempts should be made to develop competence to represent such groups by public or private agencies.
3. The results of the standards development process should be published, together with explanation of the reasons why the standards are desirable in achieving its purpose, identifying any opposition, and affording opportunity for dissenting comments.
4. An appeal by any dissatisfied, interested party should be permitted to an independent, impartial professional staff for decision which is published. The right to participate in the appeal should be open to any dissatisfied party who participated in the earlier development and to interested parties if good reason exists.
5. Organizations should recognize and acquire competence in standardization.
6. An annual report should be published evaluating the procedures and results from the public point of view.

Finally, a third function is: Standards-making bodies should serve the public interest by measuring the results of standards.

One can assess the impact of a mandatory standard by how it is enforced and whether as enforced it serves the purpose for which it is intended. Results can be measured from this point of view more readily than they can in the voluntary standards process. You will recall that compliance with voluntary standards is compelled by the marketplace, public opinion, and economic pressure, even though there is no collective agreement to exclude certain product alternatives not included in the standard. Voluntary standards also are enforced in a variety of ways through state and local laws and regulations. So voluntary standards frequently become mandatory. For this reason, it is imperative to have an adequate process in developing them privately. If the standard has been properly proposed and processed, society will tolerate, I think, certain pressures to compel compliance. Private agreements might even be construed as enforceable if the public purpose behind the voluntary standard is carefully enough developed and the process is carefully enough conducted. Some have called this result private legislation. I prefer to look at it as the development of customary standards to meet social needs that the government and the courts will recognize. For example, I cannot imagine any court holding unlawful a private agreement to abide by a standard which excludes a product that is demonstrably unsafe for the general public when Congress has not acted. Also, as I pointed out earlier, I am not persuaded that government approval of a voluntary standard in any way sanctifies its later compliance by industry. We should not confuse government participation in the prescribing or processing of standards with government approval of their enforcement.

The main difference between setting mandatory standards by government and voluntary standards by private institutions is in each of the two essentials – due process and purpose. For mandatory standards once set, the government assures compulsory compliance to serve the overriding purpose. For voluntary standards, there is no federal compulsion. Public opinion, the marketplace, local enforcement, and the public interest, as well as self-interest, assert a form of pressure, frequently to the point of compulsion, for compliance. In the one case society protects the interest of all by tolerating decentralized and private pressures. This distinction is crucial to understanding the role of the antitrust law which, simply put, is to prevent economic abuses in privately forcing compliance with a standard once prescribed.

If a standard should be but is not observed privately and the public interest suffers, we should have some more systematic way of reporting this to the Congress so that it might take appropriate steps to authorize government enforcement. Similarly, if a standard is inadequate to achieve the purpose for which it is designed, we should have some systematic way of either upgrading it privately or asking the Congress to authorize its mandatory revision. We have seen a good example of a technique for following this approach in the Fair Packaging and Labeling Act which adopted the voluntary standards process for achieving a public purpose of eliminating undue proliferation in certain consumer commodities. If industry is unable to agree on a standard or if one being adopted is insufficient or is not observed, the Secretary of Commerce is under a duty to report the matter to Congress with his recommendations.

I propose, therefore, that some responsible group should evaluate and appraise the results of private standards and recommend to the Congress mandatory action when it is necessary to protect the public interest.

One last suggestion: As you look within your own organizations, you are constantly confronted with two major problems, money and people. As you look at the matter of standards, it too is confronted with two major problems, money and people. It is in the best interest of the country, as well as the best interest of your own organization, that you support private standards-making bodies more generously than you have been inclined to do so far. The only way the public can support standards is through government action, but sole reliance on government is not the only way. The nation is best served by having a strong voluntary standards activity also. As to the second problem, people, too many organizations have for too long assigned standards functions to people on the lower levels of responsibility. It is in your best interest for top officers to give priority to standards-making. As a matter of policy, organizations also should urge their standards personnel to participate, along with consumers, in domestic and international standards activities conducted in the public interest. And we need this participation for the mandatory as well as the voluntary standards process.

I have taken considerable time in my remarks today to develop what I think is a sensible national policy for standardization. The private professional groups can strengthen their role in protecting the public interest. They can and should give immediate priority to the points I have outlined, for I do not think that government alone should be concerned with the public interest. The public airing now going on in the standards arena will undoubtedly educate the public to the value of standards in our society. Whether we have educated ourselves well enough to exercise leadership for standards in the public interest remains to be seen.

# FUNDAMENTALS OF STANDARDIZATION

*Abstract From "Industrial Standardization in Developing Countries," United Nations, 1964, Department of Economic and Social Affairs.*

Dr. J. Gaillard and Madhu S. Gokhale

**Natural and Organized Standardization**

What is a standard? Why do we need standards? A single example will answer both questions. All social and business activities throughout the world are based on a universal time standard which consists of a reference line (the Greenwich meridian) and standard units of time (the hour and its subdivisions). Without general agreement on this standard there would be utter confusion in communication and transportation, even in the same country.

Similarly, there would be confusion everywhere if there were no standards of weight and length. For many centuries such standards have been used in villages and towns, but, having grown up independently, they are all different and have given rise to much trouble. In fact, in contrast to the world time system, which is a universal standard, there are still in use, in the most highly industrialized countries, two systems of weight and length: the metric and the British. In addition, the old systems have some sentimental value because of their traditional use.

Some of these systems are very old; they grew up long ago and since then have become so much a part of the common language in the market-place, and later in more extensive trade activities (even with foreign countries), that for several reasons those using the systems are reluctant to part with them. To begin with, a change in system would force them to change their thinking about quantities. In addition, the old systems have some sentimental value because of their traditional use.

In languages, we have a similar situation. English, Spanish, and French are widely spoken throughout the world, but efforts to create a single world language have so far had no practical success: here again, it is not only the difficulty of making a change, but also the departure from an old tradition. On the other hand, the Morse code, designed as a standard system to meet the needs of a new method of communication (telegraphy), is understood by all concerned, whatever their native language. Here is a standard system that did not grow up in the course of centuries as a natural product of its surroundings, but was designed for a particular use and purpose.

We thus find that there is a distinction between standards that are the result of the growing up of a habit, custom, or tradition, which we shall, therefore, designate as "natural standards," and other standards that are the result of planning. In accordance with this distinction, we shall speak of "natural standardization" and "organized standardization."

In applying standardization in industrial problems, we have to deal with both kinds. Our systematic approach to the problems concerned, such as the establishment of standards for technical performance and quality of manufactured products, will be based on organized standardization. However, both the development of standards by this method and their introduction into practice provokes resistance to standards. Resistance to change caused by adherence to natural standards involves many human factors which call for careful consideration in the introduction and operation of organized standardization.

## The Concepts "Specification" and "Standard"

In industrial and commercial practice we find many different interpretations of these two terms. In some quarters the term specification has the definite connotation of a document used in purchasing. However, there are many other kinds of specifications. In a manufacturing industry alone we have design, material, dimensional, product, tool, and test specifications, to mention only a few. And building specifications cover an entirely different field.

Whatever may be its subject, we shall always find that a specification states one or more specific requirements that must be met. Why? The answer depends entirely on what the specifier had in mind when he drew up the specification. For example, if he wrote a material specification, he would write into it requirements of strength, or color, or elasticity, or the like, depending on the purpose for which he intended to use the material.

If a specifier finds a specification to be satisfactory in use, he may decide to use it every time he has to achieve the same purpose. This has the great advantage of saving him time and effort. His problem was solved by his writing of a specification. He can now keep the solution in readiness to be made effective whenever he wants, at short notice. The recurrent use of the specification thus becomes a "standard specification," or, briefly, a "standard."

In accordance with the above reasoning, we shall here, for the purpose of our discussion, define a standard as a "specification intended for recurrent use." At the present time there is no general agreement, even among standards engineers, on the definitions of the concepts specification and standard. However, the question is often raised, what is the difference between a specification and a standard? According to the definition just given, every standard is a specification, but not every specification is a standard. The difference lies in the intention of recurrent use.

## Formulation of a Standard

Since, as just explained, a standard is, to begin with, a specification – that is, a specific statement of requirements – the formulation of a standard is basically the writing of a specification.

A specification has in general to perform two functions. One is to tell another party what the specifier wants to have supplied to him in terms of goods or services. In this respect the specification functions as a communication. When the other party supplies these goods or services, the specifier will use the specification as a basis for determining whether what he is offered complies with the requirements written into the specification. The latter now functions as a criterion for making a distinction between things that are acceptable and things that do not comply with the specification.

In either function, as a communication or as a criterion, the specification should be so complete and clear that it can be understood by the would-be complier without his having to ask for further information. Terms like "good workmanship," "commercial finish," or "superior quality" should, therefore, be avoided. They may have a more or less definite meaning between parties regularly dealing with each other, but they are essentially unsuitable for making a distinction between things that should be accepted and those that need not. Those in charge of writing specifications, therefore, should always try to find, if at all possible, units and bases of measurement as a means of expressing specified requirements. In some cases this is not easy and may even be impossible. It will then be necessary to have recourse to judgment supported by the use of a sample or model serving as a basis of comparison.

Even where the use of measurement and hence the expression of requirements in numerical terms is possible, it is often more complicated to be specific than appears at first sight. The seemingly simple case of specifying the diameter of a cylindrical steel pin may illustrate this.

As the first step towards being specific in this case, we shall say that the pin must have a diameter of, for example, 20 millimeters. This might be held to be a specific requirement since it is expressed in a unit of measurement, the millimeter. Upon closer consideration, it is seen that the requirement is specific only with some qualifications. The steel pin is subject to expansion and contraction with changes in temperature. Therefore, the question arises at what temperature the pin shall have a diameter of 20 millimeters. There is an International Organization for Standardization (ISO) Recommendation, which has been adopted as a national standard in several countries, specifying that in cases of this kind, when no specific temperature has been mentioned, the temperature of 20° C shall apply. It is known as the standard reference temperature.

A second step towards being specific has to be taken in the form of a statement of the extent to which the diameter of the pin as measured will be permitted to deviate from the basic value of 20 millimeters. Both parties, the specifier and the would-be complier, are likely to agree that in practice it is impossible to make a pin with mathematical accuracy; for example, with a diameter of exactly 20 millimeters. Hence, all that can be expected is that the diameter of the pin will be held within certain limits. The two parties, therefore, will have to agree on the question of how far apart these limits should be. The answer will depend on two conditions: the purpose for which the pin is to be used – that is, the performance for which it is intended – and the accuracy with which the manufacturer of the pin can make it. If it is agreed that any pin will be acceptable whose diameter is within the limits 19.995 and 20.005 millimeters, this statement will represent another step towards being specific in stating the required diameter.

A third step may be necessary because of the fact that it may make a difference in what way the diameter of the pin is inspected for size. Accordingly, the method to be used may have to be specified if the specification is to give the would-be complier a complete picture of the conditions which he is expected to meet.

Similar considerations apply to the specification of a material. Here, the specifier may begin by stating that the requirements, illustrated by the examples just given, are:

1. Specification of the nature and basic value of a diameter (in the example, this is a dimension of length and its basic value is 20 millimeters);
2. Specification of the extent to which deviation from the basic value is permissible (in the example two limits were given; another way of stating this requirement would be to say that the maximum permissible deviation from the basic value is 0.005 millimeters either way);
3. Specification of the method by which compliance with the requirements mentioned under 1. and 2. shall be checked. (In example, this may concern agreement on a system of gauging used to inspect parts of this kind.)

In conclusion, it may be emphasized again that those who are writing specifications and, therefore, also those who are formulating standards, should always aim at stating the requirements in those documents, whenever possible, in terms of measurement, this being the most specific way of expressing them. In some cases a method and unit of measurement may first have to be established – a problem that may call for research.

**Standardization and Research**

A valuable by-product of organized standardization may result from the necessity to find facts before deciding on requirements to be written into a standard. This necessity may appear when, in discussing the formulation of a standard, technical specialists are found to have opposite opinions on a given point, without either party having factual evidence in support of his opinion. It may then be decided to seek advice from research.

Research and standardization are complementary functions. Without standard units of measurement, standard solutions, standard equipment, and countless other standards, no research laboratory could do its work. Its basic assignment is to find solutions for problems still to be solved. Standardization will distil from these solutions recommendations that will be readily available to those who need them in their work and these recommendations will be revised when research has made new progress.

This point is made here because there is still too much of the misunderstanding that standardization hampers progress. Organized standardization as a function that can and should be carefully controlled will keep standards flexible through timely revision.

Close cooperation between research and standardization may be found at all levels of standardization. A technical committee working on a national standardization project may refer a research problem to one of the interested parties represented on the committee, such as a technical society, a trade association, or possibly a research organization in the government. At the international level, a problem concerning automobile headlights which arose in an ISO Technical Committee some years ago was referred to the International Commission on Illumination for advice.

Actually, there are cases where the establishment of a central technological research agency in a country has clearly shown the need of the organization of the standardization work at the national level and thus has led to the creation of a National Standards Body.[1]

### Forms of Applied Standardization

One of the basic principles of organized standardization is to keep to a minimum the variety of things necessary to meet a given set of objectives. For example, if somebody plans to go into the manufacture of containers for shipping agricultural or industrial products, he should determine with the closest accuracy possible how many different types and sizes he should make. Once he has made this decision, he should determine how the total number of containers can be manufactured with the fewest possible materials, operations and processes, production machines, etc. This analysis will result in keeping down the cost of manufacture and other operations.

The need to restrict variety to the necessary minimum through organized standardization calls for three kinds of application of this principle: simplification, unification, and design of standards.

In an industry, especially an older one, where a large number of types and sizes of product have come into existence, a survey will very probably show that the variety is excessive and should be reduced. The existing situation can usually be greatly improved by the use of standardization in either of two forms: simplification or unification.

Simplification is the process of weeding out all varieties found not to be necessary or, conversely, keeping in existence, without change, only those varieties which are necessary. The opportunity for simplification

---

(1) See also the section below entitled "Bureau of Standards: Weights and Measures Administration."

should be expected to arise in any standardization project, whatever its level, where varieties of products or practices have grown up. A machine-tool builder found that his firm was using 50 different lubricating oils when 6 would be sufficient. A manufacturer of radio and television sets listed 450 different control knobs, which were reduced to 30.

# INTRODUCTION TO STANDARDIZATION

Rowen Glie

In my generation of standards engineers, I am, so far as I know, the only angry young man. The reason why I am angry you will see before I finish this paper. The reason for my being young is simply because I cannot help it – my mother brought me into this world very young and I am still a babe in the woods. I am telling you all this in order to prepare you for some unorthodox things I will say while giving you some of my definitions. These definitions are actually an introduction to the main subject of our meeting, namely standardization.

To understand standardization, we must define some concepts vital to it. These concepts are: routine, because standards establish a routine; conformity, because standards establish a requirement to conform; specification and standard, because the results of standardization are specifications and standards. And last, but not least, we must understand what is meant by simplification.

What is routine? I would define it as something that causes the new generation of engineers to see red. They take the word "routine" as an adjective, which the dictionary defines as commonplace or uninspired. It must be obvious that no engineer should be commonplace or uninspired. However, the word "routine" taken as a noun means: "regular course of procedure and/or habitual mechanical performance of an established procedure." The first defines the word "routine," the second defines rather the result of a routine procedure. It becomes obvious from this definition that routine is a blessing to any manager. Without an established routine, nobody in an organization would know what to expect from anyone else. What is managerial planning if not devising steps and procedures to be followed by personnel in order to achieve the company's aims? But, an established routine is more than a disciplinarian instrument by which the company keeps its personnel firmly in its hands. A good company routine is not the fruit of chance but the result of study, evaluation, and test. The routine tells us what things are, how things should be done, who will supply or do them. Routine therefore is a fruition of a learning process. It is utilization of company engineering and technical experiences.

Engineering routines are company know-how. They are the most important things that a company has to offer to its customers. Behind each nationally known brand name, behind each slogan – "Progress is our most important product," "You can be sure if its Westinghouse" – behind each company reputation is an effort to impress the customer with the company's know-how, i.e., with the engineering and technical routines. When a routine becomes firmly established, the work is done quasi-automatically. It is like driving a car. We truly may say that we can drive only if we do it mechanically by reflexes, without thinking and without looking up the manufacturer's manual while driving at full speed to find out what pedal is for stopping and what for accelerating.

There can be no scientific management without written routines. Only regular procedures can avoid confusion and thus reduce the amount of possible surprises which in engineering are, as a general rule, disagreeable. Only regular procedures will permit realistic prediction of time, manpower, and money required to complete a project. On military contract it will also permit the check by government inspectors and reliability engineers of the soundness and consistency of the contractor's processes and methods.

The difference between an organization firmly under control of its management and a loose agglomeration of building, people, and machines lies in availability of regular procedures, i.e., routines. Taking an extreme example, it is like a disciplined advancing army and the same army in rout, with each one for himself.

In conclusion, I would like to add that standardization is defined in many companies as: "delegating the problems that have been solved to the field of routine." This is not a very good definition of standardization but it is an important one, mostly because it expresses standardization in terms of company routine and provides you with the first clue to standardization: While not all routines are standards, all standards are routines.

Conformity – conform is defined in the dictionary thus: to bring into harmony and agreement; to be similar or identical; to be obedient or compliant or to adapt oneself to prevailing standards and customs. It seems innocent enough, but if the word "routine" makes the modern generation of engineers see red, the word "conformity" makes them see infrared. Why?

Because the modern generation of engineers was innoculated in our schools and universities with a contempt for conformity, it is in universities and engineering schools that the idea is generated that engineers, being creative people, cannot be conformists.

This shows that the modern engineering education is losing its touch with technological reality. The fact is that modern engineering must rely on conformity to a degree which would have been incredible a dozen years ago.

It is interesting to note that while there are college courses in creativity there are no courses in conformity. The very idea of such a thing makes many of us, even standards engineers, shudder.

We shall see that this ideal is not so crazy as it seems. Returning to creativity, we may ask ourselves how creativity, which is truly an inspirational concept, can be taught. How do you teach inspiration? Well, the most widely accepted method of creativity goes as follows: "Defy it! Make it different! Modify, magnify, minify, substitute, rearrange, reverse, put straight, make crooked, combine, separate, simplify, complicate!" In other words: if you cannot create – change!

These recommendations may be used as a tool to maintain an open mind and as a means to evaluate different possibilities of a design but, used by noncreative engineers who need to show that they are creative, this creativity effort becomes a passion to make changes for the sake of changes and cheap glory. This type of creativity never brings an engineering breakthrough. It only increases variability in terms of small, unnecessary variations of the same basic idea and creates havoc with military and nonmilitary logistics, interchangeability, and maintenance.

Let me give you just one example of what happens to the consumer when too much variation exists in a market. I quote from the New York Times, April 12, 1966.*

> London, April 12 – If Prime Minister Harold Wilson really wants to bring Britain into the brave new technological world, he should start learning the problems by trying to install an electric clock at 10 Downing Street – by himself.
>
> That is the conclusion of an American resident of London, who set out the other day to provide her kitchen with a clock. Her story typifies the bewilderment of foreigners here at the eccentricities of British life.
>
> She bought the clock at Peter Jones, a large and superior London department store. It was a modern electric model, about

---

*"American Makes a London Connection." Copyright ©1966 by The New York Times Company. Reprinted by permission.

the same as the kitchen clock she would have bought at Macy's.

The first difference appeared when the woman got home and opened the box. She telephoned Peter Jones.

"Ah, Madam," the salesman explained, "You must attach your own flex (cord). That allows you to choose one of the desired color and length."

She had an extension cord left by the previous tenant, and so she proceeded.

It took a frustrating 20 minutes to attach the cord to the clock. There were five tiny screws – one to remove part of the back, two underneath for the electrical connections and two to hold the cord in place.

That was the easy part of the job. The hard part was at the other end of the cord – the plug.

Every American who has lived in London has grappled with the electric plug problem, his amusement gradually giving way to irritation. There is no standard plug design.

There are plugs with three prongs in a row and plugs with three prongs in a triangle. There are plugs with two prongs. There are round prongs and square. There are large plugs and medium-sized and small and extra-small.

No plug will work except in the identical socket. And nearly every house has a half-dozen different kinds of sockets.

The plug now at the end of the American woman's flex was too small for the only electric outlet in the kitchen. And changing it was not a simple job.

Over here, putting a plug on a cord means fiddling with up to seven screws: two to take the plug apart, two to grip the cord and three to make the electrical contacts.

After the plug was changed, there was still the question of what it was to go into. The one kitchen outlet was in use – and the neighborhood electrical shop did not have a multiple adapter plug. Neither did the shop down the road.

The American managed to rig up a homemade system to plug in her clock.

It was nearly two hours since she had cheerfully unwrapped the package.

That evening she told the story to an English friend. She added that half the lighting fixtures in her house took screw-in light bulbs and half, snap-in bulbs.

"That's nothing," the Englishman said. "The other day I bought an electric fire (heater) at the local office of the Electricity Board. When it was delivered, I happened to notice on the box the words, "Direct Current." Our neighborhood, like 95 percent of London, has alternating current.

"I went back to the Electricity Board office and said they'd made a mistake and sent me a D.C. electric fire.

"The man's face went white. Without saying a word he went out of the room and ran down some stairs to the basement – I could hear him. When he came back up, he looked glazed. He said:

"They've sent me 120 of them."

Creativity should not be glorified at the expense of conformity – they go hand in hand. It is hard to realize how much creative effort is spent by people in NBS, EIA, ASA, DOD, in technical societies, in creating standards, i.e., conformity. It is also incredible that conformity needs defense.

True, conformity for conformity's sake is as baseless as creativity for creativity's sake. But, even on such an obviously highly creative project as Apollo, it is mandatory that engineers conform with NASA requirements from the format of the drawings to the NASA engineering specifications, that anything supplied on the project, be it material, component, instrument, or process, conform 100 percent with the required specifications and/or standards.

An army of thousands of quality assurance, reliability, components, application, standards engineers and inspectors spend and will spend hundreds of thousands of hours to check, to inspect, to verify, and to prove by millions of test hours and statistics – what? Conformity!

Money and time spent on the Apollo project on checking conformity stagger the imagination. Without this conformity, the most ambitious of human projects, the conquest of space, would still be science fiction and not an engineering accomplishment.

Realization of what an engineer dreamed up is possible only because other engineers created areas of conformity which are the preconditions for technological realization. Without conformity, the creative engineers would create only paper tigers at best and, because of frustration, pink elephants at worst.

*The product of standardization is conformity within strictly controlled limits.*

*Specification.* Now we ask ourselves, "What is a specification?" To answer this question we must take a broad, bird's-eye view of technology in general.

What is technology? It is man's answer to his environment – with technology we control it. We have to, since, unless we control environment, the environment will control us – influence us and our lives, and the result could be tragic. Dinosaurs, who were beings of tremendous strength, became extinct because the climate became colder and they did not discover fire, invent houses or clothes to warm themselves.

The control of an environment is not an easy problem. To understand what actually happens, three things must be realized: (1) we live in an everchanging world; (2) to survive we must control it within certain limits; and (3) to control it, we must forecast the changes in it.

## Our Everchanging World

We cannot emphasize strongly enough that we live in a constantly moving, self-transforming world, undergoing ceaseless changes. There is in the world a continuing Einsteinian transformation of matter into energy and energy into matter. Stars explode and become nova and new stars are formed.

There are ceaseless changes around us – climate changes – the firm ground we walk on changes – where there were mountains there are now seas – where there were once seas there are now mountains. There are changes in us – within about three years, all the cells in our bodies are replaced by new ones.

Things change without any human intervention. They expand and contract – "breathe" with temperature changes, they augment their volume with increase of humidity; some materials "age" – become harder just by lying on a shelf; they crystallize, decrystallize, rust, start cold flow, become brittle or soften, get ionized, deionized; they grow, they crumble, they combine or fall apart.

In engineering, we cannot accept changes unless we control or predict them. Things cannot be permitted to start to change unexpectedly as, for instance, a capacitor to crystallize immediately after it becomes charged, or a transformer's insulation to vaporize after it becomes heated up.

It is obvious that, unless an engineer can predict the changes far in advance, it would be impossible for him to use any component or material as a part of his design.

In engineering, predictions are made every day. Acceptance of an R & D order is based on our ability to predict time, manpower, and money necessary to transform a project on paper into a functioning system performing as required.

But, engineers are not prophets, not seers, they do not get visions of things to come, nor revelations as to how a project will proceed.

*But, if we cannot truly predict the future, we learn how to create it.* By manipulating materials, components, and processes, we create a chain of causes which will produce the desired effects.

Any functional diagram, design layout, or mechanical schematic shows sequential relationships between events and entities in the same time series, such that when one occurs, the other necessarily follows. However, we can be certain of our effects only if we are certain of our causes.

What are the causes that an engineer uses to get his result? Obviously they are energy – electrical, mechanical, or other – materials, components, and processes.

But we just showed that we live in an everchanging world. If, then, our causes constantly change, the effects (i.e., the results) will also change. This is not admissible since this will create doubts and uncertainty of the final result. Therefore, we must control the changes of items.

We select an item for our design because it has certain characteristics. These characteristics will be causes which insure the desired effect. Therefore, the item selected must become immune to changes insofar as we can control this.

We want the item delivered to us exactly as we see it – not worse and, often, even not better. We have then, the following problems:

1. To clarify in our minds what type of an item we need
2. To communicate the image of an item we have in mind to everybody involved, i.e., to designers, purchasing, shop, vendor, and customer, so that when we all think about the same item we will have the same image in our minds

We have then a problem of description of an item which we will use in our design. To describe an item is obviously to describe its specific characteristics.

This presents a complex problem since any description means use of words, and words immediately beget semantic interpretations. What may be white for me may be gray for you and what is good for me may be unsatisfactory to you. We can eliminate completely any problems of mutual misunderstanding if we replace words by measurements.

Now we come to the point when we have all the elements necessary for comprehension of the concept of a specification.

A specification is a written communication between interested parties describing a required item or process in terms of its peculiar or required characteristic(s) and/or performance(s).

The quality (amount) of each characteristic(s) and/or performance(s) are defined in terms of mutually acceptable methods of measurement.

The deviation from the desired quality is limited by means of tolerances assigned to each quality.

Thus a specification achieves the following:

1. It makes the item or performance "real," i.e., the same for everybody involved.
2. It insures repeatability of the item or performance.
3. It permits the possibility of predicting the behavior of the item taken alone or in combination with other items.
4. The item becomes useable in engineering design because it may be used as a cause to produce the desired effect with a good confidence level that the effect will be as predicted.

We may conclude that no one material, component, instrument, or process may be used by engineers unless there is a specification for it. *When engineers talk about an item they talk about its specifications.*

Engineers are buried up to their ears in specifications and I am not kidding: the government cannot buy anything without a specification; it has 35,000 of them. Each article in a magazine describing a material, component, or a process is actually a specification. Manufacturers' catalogs are specifications.

Not only has each item offered to engineers a specification, but the same item may have many specifications, such as military, federal, SAE, AIEEE, SME, and so on, and so on, including specifications of individual companies. "Be fruitful and multiply" applies as well to specifications as it does to engineers who create them. As usual, the explosion of population creates disorder, confusion, and trouble, and the situation becomes hopelessly unmanageable.

### Standardization

Every day there are new engineering developments, new materials, instruments, machines, processes. Probably 20 percent of items we currently use were nonexistent 10 years ago and belonged to the domain of science fiction 20 years ago. All these items are offered to engineers for use on new design. Each is described not by one, but by many specifications. In addition, the items are known by different proprietary names.

There are, for instance, AISI, ASTM, MSA, federal, and many military specifications to describe each of about 50 different types of corrosion resistant steels and, in addition, these steels are known by more than 3,000 trade names.

This, of course, does not help the designers with their selection, the purchasing agents with their buying, the stockroom with its replenishment and stocking, and the reliability engineers with their data on material failures, nor does it help the military services with their vital problems of logistics.

Transistors, the invention of which was a major technological breakthrough, are now sold in 18,000 or more variations, with each transistor having a few dozen characteristics. In addition, some of the specifications covering them presently are so general that the same specification may cover many noninterchangeable items.

The fact that there is a great variation of the same item is not a proof of creativity of the engineers who created it but is surely a pain in the neck to anybody who has to design, to maintain equipment, or repeat performance, or just use the equipment.

The situation is worsening when we consider large engineering departments with hundreds of engineers that come from many different companies. This type of situation is neither healthy for an engineering company, nor for DOD, nor, obviously, the nation.

Let me give you two more examples. I am sure they are familiar to all standards men, but may be new to the nonstandards engineers: a few years ago, an epidemic of cholera broke out in Egypt. Immediately, the United States sent syringes and Sweden, syringe needles to help fight the epidemic with serum. Unfortunately, the needles did not fit the syringes. By the time the situation was corrected, thousands had died. Had an international standard for surgical syringes and needles existed, confusion would have been eliminated.

During World War II, the campaign in Africa was nearly lost because somehow it was not realized that fasteners supplied by the United States will not fit British tanks and so the tanks could not be maintained. German General Rommel did not wait for us to bring the right screws, he went ahead with his offensive. The result of this situation was the so-called unified screw standard – a joint American, British, and Canadian managerial effort to correct a situation which could become tragic if left to itself.

When a state of confusion exists in any organization or group of people united by common links of interest, such as family, nation, industrial company, or Department of Defense, the confusion must be eliminated or the group will have its survival threatened. When rivalry or competition exists, nothing is more deadly for a company, nation, or Department of Defense than a state of confusion – common misunderstanding, unreliability, uncertainty, unpredictability. Confusion within any kind of human group or organization can be eliminated only by good management.

Standardization is the managerial action which strives to eliminate confusion and create order. A standard is simply a name for managerial control of possibilities open in technological state of art.

National standardization is a voluntary effort of self-management; it provides for better utilization of all American resources. From the point of view of national defense:

> Standards are documents that establish engineering and technical limitation and applications for items, materials, processes, methods, designs, and engineering practices. They will:
>
> 1. Limit the selection of materials, items, services, etc, in order to provide for:
>    a. Functional and physical interchangeability of parts, components, subassemblies and equipments.
>    b. Compatibility of items and equipment in their own or related systems.
> 2. Establish basic engineering terminology and codes.
> 3. Limit the variety of end-use items which can be procured for stock and issue.

As you can see, results expected from standards are exactly the same as the results expected from good management. This illustrates the major difference between a specification and a standard. A specification is just a scientifically worded description while a standard is, in addition, a result of evaluation, coordination, elimination, selection, decision making, and enforcement.

From the company point of view: A standard is rationally *selected* specification and the process of standardization is the selection of a product, or a process, or a method which will be used preferably for repeatable (routine) operations.

In practice, standardization results in establishing and keeping records of a company's routine operations and in the establishment of company routine for purchased and/or manufactured products. Here again, since any selection requires a decision, standardization is a managerial function.

**Simplification**

Last if not least, we must understand the concept of simplification.

It is so typical of the reluctance of American engineers to accept standardization that Mr. Hoover, who was one of the fathers of American standardization, chose to select another word – "simplification" – to make the American public swallow the seemingly bitter pill of standardization.

Many tried to find a difference between standardization and simplification. In reality, there is no difference. They are the same. Simplification is simply the methodology of standardization. It is merely what standards engineers do.

Simplification consists in evaluation followed by elimination of excess types, grades, sizes, finishes, etc., or superfluous or unnecessary methods, and in selection of the preferred product, process, or method and enforcing its use. The result of any simplification is a standard.

In conclusion, now we can redefine in one minute the fundamentals of standardization.

*Specification* is: A scientific objective description of a technologically useable item – a component, material, process.

*Standard* is: A specification selected by management as realistically describing the preferred component, material, process.

*Standardization* is: Management effort to promote an efficient use by company personnel of materials, components, processes available in technology, as a result of engineering creativity. The effort consists in rational selection of applicable specifications and making use of them in the company routine.

I realize, of course, that the students in standardization will be satisfied with the formulation of the terms specification and standards as given above. A sophisticated standards engineer surely will ask me, as Socrates would, to define my terms. What is meant by "Scientific objective descriptions" and "realistically describing the preferred components"? To satisfy the standards engineer here are more advanced versions of definitions.

*Specification* for an item is: A description of an item (material, component, equipment) by enumerating its particular properties, each one defined in terms of its measurement and/or with a specific test; with the result that, when the item is manufactured or purchased in conformance with the specification, the user will have a replica of the item which the specification writer "had in mind when he wrote the specification."

*Process Specification*: A process specification is a listing of a set of measureable sequential operations with adequately specified materials and equipment. Conformance to these operation materials and equipment insures achievement of results forecasted by the process specification writer.

*Standard*: Standards are courses of action selected by organizational management from many alternatives to which the managed must conform in order to achieve the desired, forecasted results with the least expense of cost and time. These courses of action are actually written specifications which, after being selected for use in the organization, become organizational standards.

And finally, I would like to say this: No engineer should be ignorant of standardization and the necessity of a great amount of controlled conformity for a scientifically and economically managed company.

An engineer must be well informed. Too much innocence is not becoming to a grown man. Such engineers remind me of a story of a 6-year old boy and his father whose car had a flat tire just before a wall surrounding a nudist colony. While the father started to change the tire, the boy ran out of the car, found a hole in the wall, and looked into it.

"What do you see?" asked the father.

"Only people," said the boy.

"Only people! Are they men or women?"

"I don't know. They have no clothes on," said the boy.

# DEFINITIONS OF TERMS USED IN STANDARDIZATION

International Standards Organization (ISO)

## Standardization

Standardization is the process of formulating and applying rules for an orderly approach to a specific activity for the benefit and with the cooperation of all concerned, and in particular for the promotion of optimum overall economy taking due account of functional conditions and safety requirements.

It is based on the consolidated results of science, technique, and experience. It determines not only the basis for the present but also for future development, and it should keep pace with progress.

Some particular applications are:

1. Units of measurement
2. Terminology and symbolic representation
3. Products and processes (definition and selection of characteristics of products, testing and measuring methods, specification of characteristics of products for defining their quality, regulation of variety, interchangeability, etc.)
4. Safety of persons and goods.

## Standard

A standard is the result of a particular standardization effort, approved by a recognized authority. It may take the form of:

1. A document containing a set of conditions to be fulfilled (in French, *norme*).

2. A fundamental unit or physical constant. Examples: ampere, absolute zero (Kelvin). (In French, *etalon*).

3. An object for physical comparison. Example: meter. (In French, *etalon.)*

### Specification

A specification is a concise statement of a set of requirements to be satisfied by a product, a material, or a process and indicating, whenever appropriate, the procedure by means of which it may be determined whether the requirements given are satisfied. Note that:

1. A specification may be a standard, a part of a standard, or independent of a standard.
2. As far as practicable, it is desirable that the requirements are expressed numerically in terms of appropriate units, together with their limits.

### Simplification

Simplification (variety reduction) is a form of standardization consisting of the reduction of the number of types of products within a definite range to that number which is adequate to meet prevailing needs at a given time.

### Unification

Unification is a form of standardization consisting of combining into one, two, or more specifications in such a way that the products obtained are interchangeable in use.

### Specialization

This term is, properly speaking, not within the sphere of standardization if it is used in the following meaning:

> "Specialization is a process by which particular production units concentrate on the manufacture of a limited number of kinds of products."

### Functional Interchangeability

Functional interchangeability is achieved when those characteristics of the finished product which affect its operation have been standardized to the necessary degree of accuracy. This frequently includes characteristics other than linear dimensions.

### Dimensional Interchangeability

Dimensional interchangeability is a partial aspect of functional interchangeability and is achieved when the linear dimensions of two finished products are sufficiently close to one another to ensure only interchangeable replacement.

# THROUGH HISTORY WITH STANDARDS

*An illustrated Text Book for*

YOUNG  & OLD

PRESENTED BY THE

AMERICAN STANDARDS ASSOCIATION

70 East 45th Street, New York 17, N.Y.

## How a King's Arm Became an Early Standard of Measurement

One of the first known attempts to set up a standard in the Western world took place in the year 1120, at the first Crusades.

The sponsor was King Henry I of England, whom the Encyclopedia Britannica describes as a well-educated monarch. Henry ordered that the ell, the ancient yard, should be the exact length of his arm, and commanded that distance henceforth to be the standard unit of comparison of lengths throughout his kingdom.

The ell, 45 inches in length, is still used in a few areas for measuring cloth.

*What are thou? Have not I*
*An arm as big as thine?*
SHAKESPEARE, Cymbeline

## How Venice Built Ships in the 15th Century

Venice, the world's first modern state, apparently created the first true assembly line.

Throughout the 15th and 16th centuries, a Venetian water basin called the Arsenal, which employed 16,000 men and manufactured everything from cannon to nails, was the world's largest industrial plant. There Venice built a standardized trading vessel which could be quickly converted into a galley of war.

When finished, the galley was towed slowly down a canal, and, as it passed, workmen reached out of windows on either side and loaded on its equipment. When the vessel reached the end of the street, everything necessary was on board, including cordage, oars, armament, food, and a full complement of men.

"In this manner," wrote a Spanish visitor in 1436, "there came out ten galleys fully armed, between the hours of three and nine." In 1570, during the war with the Turks, the Arsenal turned out a hundred fully outfitted galleys in a hundred days.

*What the devil did he want in that galley?*
MOLIERE, Les Fourberies de Scapin

## Why a Standard Brick Became Compulsory in 17th Century Boston

A term in the stocks with a pile of out-size bricks to contemplate – that could have been the penalty of early builders back in 1689 when Boston was destroyed by fire. It was a civic crime for brickmakers to make bricks other than 9in. X 4in. X 4in. in size. City fathers recognized standardization as essential to the most economic and speedy rebuilding of the city.

Even 275 years ago the bricks were only one inch off the idea of the Modular Method which is now being rapidly advanced in the building industry. The activities of Project A62, organized by the American Standards Association in 1939, are spearheading this advance.

The Modular Method is the standardization of all building materials on a uniform basis of dimensioning. The architect figures everything that goes into a building on the basis of a standard module – multiples of four inches – in all directions.

Thus lumber, bricks, tile, concrete blocks, wallboard, flue linings, and framing fit in place without alteration and leave space for doors, windows, cabinets, and other pieces to slide into place.

The four-inch module was chosen because (1) it is large enough for manufacturers to reduce the number of stock sizes and still satisfy consumer demand; (2) it is small enough for ample freedom in architectural design and for flexibility in equipment layout; (3) it coincides with the dimensions of a great many building materials already standardized and is applicable to present construction practices; (4) it is a unit measurement with which architects, builders, and carpenters already are familiar; (5) it approximates 10 centimeters (3.9 inches), the basis proposed by metric system countries working on the same problem.

The module means the elimination of many odd sizes and thus permits smaller inventory for manufacturer. It means smaller inventories and speedier service for dealers. It permits architects to spend less time in drafting and more time on improved design. Easier estimating, more efficient methods on the job, less construction time are possible for the contractor. When the modular method is used, everything fits. There is no needless, wasteful hand-cutting at the building site.

*To be left alone*
*And face to face with my own crime, had been*
*Just retribution.*

LONGFELLOW, Masque of Pandora

## How Mass Production Began with a Contract for Muskets

Eli Whitney has sometimes been called "the Father of Standardization," and this is why.

At the end of the 18th century, under the authorization of the Fifth Congress, Vice-President Thomas Jefferson gave the inventor of the cotton gin a contract to produce 10,000 muskets. Whitney built a small factory in New Haven and started to work.

Up to that time, the traditional manufacturing method was one in which each man, highly skilled, produced by himself a particular part of the finished job. Whitney took the unprecedented step of dividing his work into steps and putting groups to work on each step in a simplified operation. Each part, copied from a model musket, was interchangeable with its standardized counterpart in the other muskets.

Whitney introduced three new aids in this work: drilling by templates (patterns), filing by jigs (guides), and milling irregular forms. He did not, however, introduce gages to inspect and control the accuracy of the product, but rather depended on keeping his tools sufficiently accurate to make the parts interchangeable.

Tooling up required many months, and Congress became impatient. At a critical stage, Whitney appeared in Washington before an assemblage of government experts and Congressmen. He laid out a pile of parts and assembled ten muskets by picking standard parts at random – an unheard-of feat. Congress was convinced, and thus mass production was started on its way in this country.

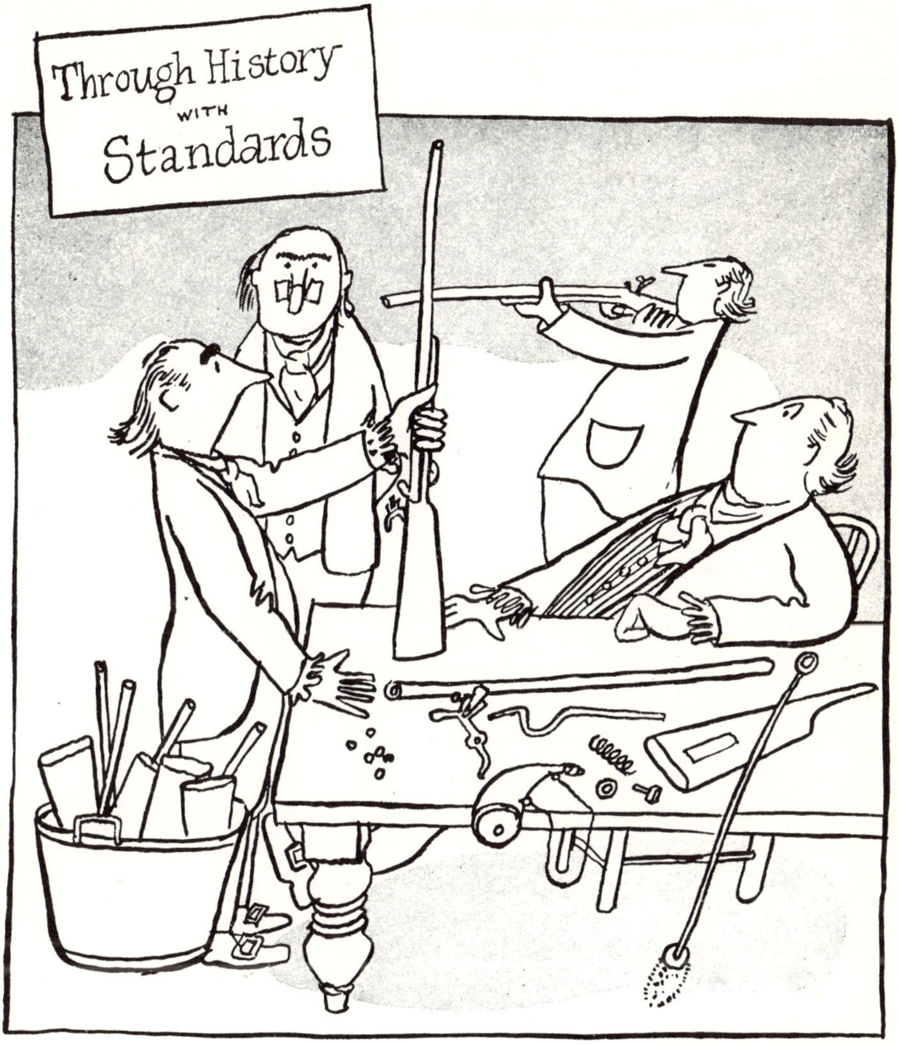


*As lyke as one pease is to another.*
LYLY, Euphues

**How a Common Household Problem Inspired a 19th Century Toolmaker**

The candle that doesn't fit into the candle-holder has long been an annoyance to householders. This problem was taken up early in his career by Joseph Whitworth (1803-1887), great Manchester engineer and toolmaker – later, Sir Joseph Whitworth – who was probably the first British engineer to preach standardization as a means of obtaining interchangeability. Whitworth illustrated his thesis by pointing out a need that everyone could feel and understand – the need of standard sizes for candle-butts and candlesticks.

Whitworth later developed a system of standard gages that was widely used in British industry. In 1841, in a paper read before the Institution of Civil Engineers, he urged the adoption in Britain of a uniform system of screw threads in place of the various different pitches and dimensions then used. His thread design became known as the Whitworth Thread and was adopted both in Britain and on the continent.

In 1948, after more than thirty years of negotiation, America, Britain, and Canada agreed to merge the Whitworth Thread and the American Standard Thread into a Unified System of Screw Threads. It is the thread now used in virtually all military equipment and in most of the civilian goods produced by the three countries.

Candle-butts and candlesticks have been standardized to some degree, but time and technology have made this problem somewhat less acute than it was when Whitworth studied it in 1830.

*The Great Adjustment is taking place.*
HARDY, There Seemed a Strangeness

**How a Standard for Steam Boilers Brought New Safety to an Old Hazard**

From the earliest practical applications of steam for prime movers in the 18th century, boiler explosions were a serious cause of loss of life and damage to property. Steamboat boiler explosions were especially common, comparing in frequency and severity with the worst catastrophes of today. On April 27, 1865, for instance, the steamboat Sultana, plying the Mississippi River near Memphis, exploded with the loss of 1450 lives.

In 1911 the Council of the American Society of Mechanical Engineers appointed a Committee to Formulate Standard Specifications for the Construction of Steam Boilers and Other Pressure Vessels and Their Care in Service (known since 1914 as the Boiler Code Committee). It issued the first ASME Boiler Code three years later.

The makers of this code set out to create a set of rules for the construction and care of steam boilers that would be adopted by all the states of the union. Today, 28 states and territories, eight provinces of Canada, Australia, Mexico City, and a number of other countries and cities have adopted one or more sections of the Code. Through its extensive use and uniform administration, the Code is now in every sense of the word a national standard.

The 1952 Code, as revised and enlarged, has nine sections totaling some 1,000 pages. It is administered by a Committee of 24 members and an Executive Committee of eight, made up of men of special training and experience representing designers, fabricators, suppliers of materials, operators, and consultants. Some three dozen subcommittees serve under the Code Committee.

Since 1916 the record of serious boiler explosions has been exceptionally low, despite the great increase in the number and size of boilers. For this, the Boiler Code Committee has been in a large measure responsible. In a citation to the Committee in September, 1952, the American Standards Association stated: "Probably no other single standard in America has done more for national safety than the ASME Boiler Code."

# THROUGH HISTORY with STANDARDS.

*It shall be a vexation only to understand the report.*
ISAIAH, XXVIII

## How the Railroads Standardized Their Track Gages

Standards engineers agree that probably the most important group standard ever developed in the U. S. was the railroad's standard track gage and system of interchangeable brakes and couplings. These standards make possible the interchangeability of rolling stock among virtually all roads throughout the nation.

Most of the early American railroads had their own gages, the changeover to the present standard gage being made about the time of the Civil War. At one time President Lincoln issued an order that the gage should be five feet. This never became effective, and after many consultations between the roads the present gage of 4 feet 8½ inches became universal.

This standard gage is used in Britain, the U. S., Canada, and most of Continental Europe. India and most of the important railroads in South America use 5 feet 6 inches. South Africa uses 3 feet 6 inches. Australia has never adopted a standard gage, and when cargoes are transferred they have either to be reloaded or special undertrucks have to be used.

*Around the ancient track marched, rank on rank,*
*The army of unalterable law.*

MEREDITH, Lucifer in Starlight

## How the Railroads Standardized Measurement of Stresses

One of the problems that used to plague engineers was why an iron structural member would give way under a load that was clearly less than the load that tests had shown it could endure. This problem came to the fore in the first half of the nineteenth century, when it became necessary to establish standards for determining in advance the amount of service that could be expected from the wrought iron bridges that were then beginning to replace masonry bridges.

Thanks to the work of Eaton Hodgkinson and a commission appointed to study the application of iron to railway structures, it was discovered that the ultimate static load is not the only factor to be considered in determining the strength of a structure. It was learned that a load need not be the maximum load, as determined by test, to cause failure – if it were repeated often enough. It was also learned that deterioration of the material was not the determining factor in failure.

Experiments conducted by Wohler established that the range through which a stress is varied has direct correlation with speed and failure. That is, under the practical conditions of use of railway structures, for instance, load variations imposed by passing trains and the hammer effect of steam locomotives were more conducive to failure than would have been the application of a constant load.

The effects just discussed are now widely recognized and are incorporated in standards issued by the Association of American Railroads and the Association of State Highway Officials.

# Through History with Standards

*Light burdens, long borne, grow heavy.*
HERBERT, Jacula Prudentum

## How a French Balloon Ace Paved the Way for Simplified Design

In the 1870's, Charles Renard was known in France chiefly as a hero of the sky, into which he had ventured several times on experimental balloon ascensions. Today, in the first decade of the Jet Age, he has won world-wide recognition for a scientific principle he developed long before the time of powered flight.

As a balloonist and as secretary of a French "commission on aerial communications," Colonel Renard faced a comparatively new set of problems in devising specifications for the gear required for military balloons. He found, for instance, that the French Army had adopted no fewer than 425 different sizes of cordage for mooring balloons.

In trying to determine the optimum diameters for the cordage. Renard hit upon a new principle. The different sizes, he felt, should increase from size to size in a geometric ratio based on a progression of weights per meter of cordage.

Renard used this principle of geometric progression for reducing cordage to 17 standard sizes, and for standardizing toggles, eyelets, and pulleys. He then extended his principle to develop a series of "preferred numbers," still known in France as the "Renard series."

*We have seen mighty men ballooning high,*
*And in another moment bump the ground.*
MEREDITH, The Warning

**How Men Met in the First International Conference on Standards**

The first full-scale formal conference on international standards was held almost 80 years ago. After five years of preliminary work, the "plenipotentiaries of most of the leading occidental nations," including the U. S., met in Paris in the spring of 1875 to attend an International Conference on Weights and Measures. Of the 19 nations present, 17 signed a convention which provided for the establishment and maintenance of a permanent International Bureau of Weights and Measures to be situated near Paris. The French Government set aside a plot of ground in the Park of St. Cloud for the Bureau and declared it neutral territory.

The latest international meeting of standards was the Second Triennial Plenary Session of the International Organization for Standardization (ISO) held June 9-21, 1952, in New York. Thirty-four national standards bodies belong to the ISO, the American Standards Association representing the United States.

*Somehow, somewhere, meet we must.*
WHITTIER, Snowbound

## How Industry Reduced Occupational Injuries and Deaths Through National Safety Codes

In this century American industry has achieved a remarkable and impressive record in the field of occupational safety.

One reason for this good record is the emphasis laid on safety standards. Many groups are cooperating in a drive to make these standards truly national in scope, and they are using the machinery of the American Standards Association to do it. A typical example is the American Standard Safety Code for the Use, Care, and Protection of Grinding Wheels.

Just after World War I, 14 states had laws regulating the use of abrasive wheels. No two of these regulations were alike. And insurance company provisions for safe use of grinding wheels differed from those in the states.

In 1922 a safety code for grinding wheels was drawn up and approved as an American Standard under the joint sponsorship of the Grinding Wheel Manufacturers Association and the International Association of Industrial Accident Boards and Commissions. The code contained complete rules and specifications for safety in the storage, handling, mounting, and operation of wheels; for protection flanges, hoods, chucks, and revolving cup guards; for operating speeds; and for materials suitable for use in both guards and wheels.

Having national status, the code was adopted by more and more regulatory bodies. Today the revised American Standard Safety Code for the Use, Care, and Protection of Abrasive Wheels is sponsored by the Grinding Wheel Institute and the International Association of Governmental Labor Officials. It has been adopted by most state departments of labor and industry, factory inspection departments, and departments of safety and health. The 36 states in which the code prevails cover practically all the industrialized areas of the country.

The manufacturers of abrasive products are no longer hampered by lack of uniformity in state or local regulations. The buyers of the products receive a better product at a cheaper price. The workman is protected by the best safety specifications that experts can devise.

More than 160 such American Safety Standards are in daily use throughout the nation. They cover a wide variety of fields and include such standards as those for elevators, building exits, wooden ladders, window cleaning, mechanical refrigeration, power presses, pressure piping, power trucks, concentrations of toxic dusts and gases, protective clothing, drinking fountains, traffic signals, circus tents, outdoor signs, X-ray machines, and welding.

*There is one safeguard known generally to the wise.*
DEMOSTHENES, Philippic 2

**How Standardized Fire Equipment Reduced Fire Losses**

Lack of interchangeable fire-fighting equipment cost citizens of Baltimore millions of dollars when their city burned in 1904. A fire broke out on Sunday, February 7, and when it got beyond control the local fire department called on Washington, New York, and Philadelphia for help. Special trains rushed apparatus from these three cities on cleared tracks. Then the equipment stood idly by while 150 acres of the old business section was destroyed. Their hose couplings would not fit the Baltimore hydrants.

Twenty-three years later, in 1927, fire-fighting equipment from 20 neighboring towns helped save Fall River, Mass., from total destruction in another great fire. The difference was in standardization. All equipment from the 20 communities worked interchangeably. The hydrant and hose connections had been standardized.

In the intervening years the early work of the National Fire Protection Association in this field had been developed into an American standard for this fire-fighting equipment under the leadership of the National Board of Fire Underwriters, the American Water Works Association, and the American Society of Mechanical Engineers. That standard, which was reviewed and reaffirmed in early 1953, is in very wide use in the United States. The Office of Civilian Defense has recognized the importance of interchangeable fire-hose couplings in case of enemy air attack and had adopted the American Standard screw thread for such emergency equipment.

*While I was musing the fire burned.*
PSALMS, XXXIX

## How a Young Industry Proved Its Parts Were Interchangeable

A dramatic and convincing demonstration was given in London in 1908 of the rapid progress toward standardization that the United States had even then achieved in the manufacture of its automobiles.

In February and March of that year, Henry M. Leland chose three one-cylinder Cadillac cars at random from the warehouse of the agency in London. The cars were taken to the Brooklands Race Track and there were completely dismantled. A control commission of the Royal Automobile Club of England scrambled the parts into three piles of 724 each and then replaced 89 of the parts with new parts from stock.

Three new cars were reassembled from the piles of standardized parts without hand fitting. The only tools allowed were wrenches and screwdrivers; files and emery cloth were forbidden. The cars were driven 500 miles over the Brooklands Track with only one minor adjustment.

*All are but parts of one stupendous whole.*
POPE, Essay on Man

### What a National Standard Did for Electric Lamps

Today few articles of commerce represent so high a degree of standardization as electric lamps and their accessories. There is now complete mechanical interchangeability from coast to coast. Some 40 dimensional standards for incandescent lamps and about 30 standards for fluorescent lamps have been approved.

This standardization was worked out by the cooperative effort of the electrical manufacturers, the power companies, the fire insurance companies, and others. It is a small part of a total standardization program that probably affects and benefits more people than any other. Two national electrical standards codes encompass virtually all phases of electrical transmission and use.

The National Electrical Code gives detailed requirements for the installation of electric wiring and equipment for lighting, heating, and power in public and private premises. It was formulated under the sponsorship of the National Fire Protection Association and was adopted as a standard by the National Board of Fire Underwriters. First published in 1897, it has been reviewed and revised many times. It became an American Standard in 1919. Hundreds of municipalities use the code in some form of legal enactment. Insurance companies use it in giving advice to their assured.

The National Electrical Safety Code provides the rules and regulations which insure the installation, maintenance, and safe operation of electric lines and equipment, including radio installations. It was formulated under the sponsorship of the National Bureau of Standards in 1913 and was approved as an American Standard in 1932. It has been adopted in whole or as a guide to practice in almost all of those states that have rules on electrical installations.

This standardization program has created a national market and has thus permitted mass production, distribution, and use of electric lamps and appliances. The result has been better convenience, greater safety, and cheaper prices. An electric lamp cost $1.75 in 1907; it costs less than 15 cents today – and lasts a lot longer.

*And teach me how*
*To name the bigger light, and how the less,*
*That burn by day and night.*

SHAKESPEARE, Tempest

## How the Movies Built an International Market for Their Films

In the early years of the motion picture industry, stoppages and imperfections in films were common. They were caused by irregular perforations and perforation spacings in the film, by non-standard motion picture machinery, and by non-standardized methods of handling film and equipment. In 1916 the Society of Motion Picture Engineers cited such stoppages as a main cause for the lack of popularity of motion pictures among large numbers of people.

Standards have completely eliminated these difficulties. The industry is now highly standardized in processes, procedures, and product, not only in this country but internationally. After the American Standard for the location of the sound track on 16-mm film was adopted internationally in the 1930's, many other motion picture standards were internationalized. The ASA Sectional Committee on Standards for Motion Pictures, PH22, made up of 32 organizations, has developed more than 60 American Standards, and many of these have found world-wide acceptance.

Today, motion picture film from one country can be projected in the standard equipment of any other country in the world.

*A weak and niggardly projection.*
SHAKESPEARE, Henry V

## Why Traffic Signals Were a Great Hazard Before Their Standardization

As late as 1927 a color-blind autoist stood as good a chance as anyone of interpreting traffic signals. For lack of national standardization, he could choose between purple, orange, blue, green, yellow, and red lights for his directions as he drove from state to state. Green, for instance, was "Stop" in some states and "Go" in others. Red, instead of yellow, was the caution signal in New York City.

And a motorist would have to stop his modern low-vision car a block back from the next traffic light in order to see it. Early signal towers often were as high as the second story on business buildings.

These problems have been solved in two national codes – one on colors for traffic signals, and the other on street signs, signals, and markings.

A national code for colors for traffic signals was first formed in 1927 through work of the American Association of State Highway Officials, the National Bureau of Standards, and the National Safety Council. It recommended green for "Go" and red for "Stop," now virtually a world standard.

The national code for street traffic signs, signals, and markings was first started in 1927 as an American Standard under the American Engineering Council. In 1935 a joint committee of the American Association of State Highway Officials and the National Conference on Street and Highway Safety drew up a Manual on Uniform Traffic Control Devices based on the prior work of the ASA project. It was approved as an American Standard. A new edition of the manual was begun in 1944 and published in 1948, containing significant revisions to meet present-day traffic conditions. Its recommendations for height of signals and design and color of signs are almost universally followed in this country.

*The majority of men slip into their graves without having encountered on their way any signal catastrophe.*

ALEXANDER SMITH, Dreamthorp

## Through History with Standards

Because of modesty, one great event in the history of standards has not been pictured in this series of cartoons.

In 1918 five of the country's leading engineering societies came together to form what is now the American Standards Association. They were soon joined by three Departments of the Federal government.

ASA's mission was to

– simplify development of engineering, industrial, consumer, and safety standards;

– eliminate duplication, overlapping, and variations of standards activities among the hundreds of standards bodies in the country;

– weld conflicting standards into a single, generally accepted national standard, designated "American Standard"

– promote knowledge and use of standards;

– serve as a national clearinghouse for information on all standards in this country and abroad.

Today ASA is a federation of more than 100 technical societies, trade associations, professional groups, and consumer organizations, plus 2,300 companies. Its mission is still the same – to help expand the voluntary standards movement in America into a truly national program by serving as a national standards clearinghouse.

Standards are industry's Esperanto – industry's common language of performance and acceptability. ASA does not make standards; it makes standards possible. More than 1,300 American Standards are now in use throughout the United States. These cover a wide range of fields, from engineering, textiles, and photography to petroleum products, household appliances, and industrial safety.

If your company and your association are now members of ASA, we ask for your continued good will and support. If you are interested in the aims of ASA for the first time, we invite you to learn more about us, to participate in the activities of our standards committees, and to make our work more successful by your support. Write to USASI, 10 East 40th Street, New York 16, N. Y. (Formerly the American Standards Assoc.)

# Through History with Standards

# Part 2

# STANDARDS ENGINEER

*What is a standards engineer? What are the qualifications necessary to fill the job? If sometime in the future we will have the curriculum for a standards engineer, what subjects could it include? And, if a standards engineer does not become a standards supervisor, where can he go? These are the questions that are covered in this section.*

*R.G.*

# DESCRIPTION OF A STANDARDS ENGINEER

J. V. Panek & C. J. Eiwen

It may seem presumptious to attempt to answer the question "What are the functions of a standards engineer?" in the periodical of the Standards Engineers Society. Yet a survey of Society members conducted by a committee of SES Washington Section revealed a wide difference of opinion. The purpose of the survey was to determine how industry defined the position of a standards engineer. Of a total of 123 inquiries, 38 replies were received from engineers in 31 companies. The results have been most interesting and, in some instances, rather startling.

**Nature of Sample**

Before outlining the findings of the survey, it should be noted that the vast majority of the 38 replies were from large companies. This may have been because small organizations do not define positions as closely as large organizations do. Or it may be that small companies combine a number of functions in one individual and therefore find it more difficult to answer an inquiry regarding an individual position description. Notwithstanding the bias, it is believed the results are illustrative of the general situation.

Individual replies providing valuable descriptive information came from concerns including American Bosch Arma Corporation, the American Machine and Foundry Company, the Avco Corporation, Burroughs Corporation, Carrier Corporation, Chrysler Corporation, Convair, the General Electric Company, Hughes Aircraft Company, International Business Machines Corporation, Link Division of General Precision Equipment Corporation, Manning, Maxwell, and Moore, Incorporated, Minneapolis-Honeywell Regulator Company, Shell Oil Company, Sylvania Electric Products, Incorprated, and Whirlpool Corporation.

Of equal interest if not of more startling nature is that a substantial number of replies stated that their companies had no position description for a standards engineer. These companies included Bell Telephone Laboratories, Inc., Continental Can Company, Giddings and Lewis, International Telephone and Telegraph Laboratories, North American Aviation, Inc., Radio Corporation of America, Rockwell Manufacturing Company, Sperry Rand Corporation, Westinghouse Electric Corporation, and Wheeling Steel Corporation. It is obvious that many substantial firms do not give recognition to standards engineering as a distinct and individual function. This is not to say that standardization is not recognized in these firms. Many of the replies qualified their statement on the lack of a standards engineer description, adding that standardization functions were carried out by individuals with such titles as components engineer, materials engineer, applications engineer, specifications engineer, electrical engineer, mechanical engineer, engineering specialist, and engineering coordinator.

One negative reply from the manager of engineering operations at a large industrial establishment was quite frank in giving one reason why his company had no position entitled standards engineer. His reply included the following paragraph:

> "At the present time we do not have a job description for Standards Engineer or similar classification. We use the Design Engineer classification for our Standards people. In the past we did have a Standards Engineer classification, but found it necessary to use Design Engineer Classification at times because it carried a higher salary rate. This eventually led to the banishment of the Standards Engineer classification."

It is apparent from the foregoing comment, and from a number of similar remarks, that many firms are still of the opinion that standards engineering is a subordinate function which may be delegated to other classes of engineers as an additional or supplementary duty. It will require strong and united effort to change this situation. Yet it is certainly not an insurmountable obstacle. Not too long ago a physicist was a rare individual on the industrial scene, any tasks relating to physics being shunted to a mechanical or electrical engineer. And somewhat further back, the electronics engineer was a relatively strange and unknown professional.

### Functions

But getting back to the standards engineer, what does he do? The survey indicated that his principal functions are as follows:

1. Develops and administers the standardization program

2. Prepares, coordinates, establishes, and maintains company standards

3. Assists in the development and maintenance of industrial standards

4. Serves as a consultant to other engineering divisions and, as a corollary to this, administers the collection and maintenance of library of standards, specifications, and other technical information

These are the main functions which emerge from a review of a variety of position descriptions. The wordings used varied widely. For example, as regards the standardization program, functions included: establish, set up, initiate, supervise, develop, maintain, administer, participate, and interpret. The usage was even more varied in regard to standards development, where the standards engineer might: study, develop, establish, revise, review, promulgate, assist, guide, approve, supervise, maintain, publish, coordinate, promote, write, recommend, formulate, investigate, compile, and propose.

### Place in Organization

Obviously, the place of the standards engineer within an organization is governed by the general organizational arrangement and the specific functions he is authorized to perform. It is difficult to define specifically the organizational position occupied by an individual unless the complete company organization is known. No attempt was made to secure these data. However, the title of the immediate supervisor provides some indication of relative status. These titles, as indicated by the survey, included the supervisor of engineering services, the applications and evaluation manager, the chief engineer for processes, the manager of standards and procedures, and the manager of technical services. The results appear to indicate the standards engineer is usually located somewhere within the engineering department rather than in an independent department. Management support, if it exists, is through channels rather than directly, as is often cited as an essential feature of a dynamic standardization program.

### Documentation Responsibility

A summary of the documents for which the standards engineer is responsible was revealed as a secondary result. The survey disclosed the following groupings: engineering standards and standard drawings for components and parts design techniques, drafting practices, engineering routines, installation procedures, manufacturing processes, materials products, and test methods, specifications for components and parts, equipment, manufacturing processes, materials, and test procedures; and design and drafting manuals. Naturally, this listing does not include all documents included in the document library which may be supervised by the standards engineer.

### Other Features

A number of features of the standards engineer's work were stressed in many of the descriptions. The first is that he serves as a consultant to engineering sections, to development and production sections, and to manufacturing sections. His function here is described as interpreting standards and specifications as well as providing factual information. Second, as a corollary to the consultant function, he is required to keep abreast of new developments in materials, technical products, procedures, processes, and systems. Thus he is expected not only to provide information, but also to offer advice on new items and techniques. The third feature is that he meets and cooperates with people in many other departments, with manufacturing personnel, with vendors, and with engineers from other companies. Although any company engineer might be expected to perform these duties, the emphasis in the standards engineer description would indicate that these are significant features of his work.

This then is the picture of the standards engineer and his duties as gleaned from the survey made by the SES Washington Section. While 38 replies are not sufficient to draw final conclusions, it is believed that they were reasonably representative of current conditions.

# JOB DESCRIPTION FOR STANDARDS ENGINEER

Everett Woerter

Any modern program of employee selection, placement, training, up-grading, and compensation requires adequate job descriptions. These must provide detailed information on levels of skill, knowledge, responsibility, working conditions, and special qualifications required. Job descriptions may vary in complexity from a single paragraph describing main duties and responsibilities to formal multiple-page descriptions covering every aspect of the job.

From your company's standpoint, good job descriptions provide the personnel department with information which enables them to effectively screen applicants. They provide information concerning training and experience required for the job. They show the relationship of one job to another and thus set requirements for promotion. They provide data for use in determining base rates for jobs.

Familiarity with job requirements and job analysis and evaluation is necessary before attempting the preparation of a job description. The three subjects are closely related and, to an extent, each has an effect on the other. Extensive research revealed the many complexities of standardization and the obligations of the standards engineer. The attributes and responsibilities of a standards engineer are so numerous that it is hard to conceive that any individual could satisfactorily fulfill the requirements.

A review of various job descriptions revealed that the job description is not really used in the preliminary screening or interview by the personnel department. By definition, the job description covers the job responsibilities or duties rather than the job qualifications. The personnel department must be supplied with information concerning the educational requirements, experience, skill, and personal qualities necessary to fill the job.

**Requirements**

A few points are basic. The standards engineer in today's industrial structure must be a man with a basic technical education, preferably an engineer. He must possess such qualities as initiative, imagination, foresight, enthusiasm and the ability to take responsibility and work independently.

Besides having a sound engineering background and a thorough understanding of manufacturing methods, he should have an intimate knowledge of engineering design practice and be familiar with all phases of engineering drawings.

Since the standards engineer must be capable of working with a minimum of supervision, he must possess initiative and be a self-starter. Obviously, he must be able to adapt himself to the unusual aspects of the job and display resourcefulness in developing ways to accomplish desired results.

Often there is active opposition to standardization. In such an event, the standards engineer must have the courage and disposition to persevere in spite of difficulties, discouragement, or indifference. Under these conditions, he must be adept in the art of "soft sell." His ability to get along with people will be put to the test day in and day out. Standardization is a result of teamwork and, accordingly, he should let it be known that he is available and willing to cooperate with all concerned.

The prospective standards engineer must have a clear understanding of the cooperative nature of standardization work and be able to gain the confidence and respect of his associates in other departments.

He must be enthusiastic in promoting the advantages of standardization. An orderly mind and the ability to put problems in their proper perspective is extremely important. It is also important that he not be frightened by a wealth of information. He must integrate, condense, and simplify such information, and, in any event, avoid adding complications of his own.

One of the replies to a questionnaire submitted to various companies suggested that the standards engineer have the "curiosity of a little kitten." Very true, he can't take anything for granted and he must investigate fully before making statements or arriving at any conclusions.

It is necessary that the standards engineer be capable of expressing himself both orally and in writing. He must be cautious in his method of presentation; in particular, it is advisable to present the situation in such a way that the departments concerned will not feel that the decisions they made in the period before standardization are being criticized.

It should be clear at this point that the qualifications for a standards engineer are so numerous and so stringent that compromises are often necessary. It becomes essential to capitalize on an individual's better points and develop the others by training and experience. To summarize, the standards engineer ought to be three things: a practical diplomat, a good all-around engineer, and a man with more than ordinary skill in language. The official publication for the German Standards Institute had the proper definition for the standards engineer, short and to the point, "Standards Engineer – Wonder Boy."

In carrying out his assignments, the standards engineer will meet and work with people in many other departments of his company, with vendor and supplier representatives, with engineers from other companies, and a host of others. It has been stated that standardization is more an exercise in human relations and a battle for understanding than it is a technical program.

The attitude of the standards engineer must be optimistic, cautious, and pragmatic. He is optimistic about his ability to overcome his age-old adversaries, tradition and indifference. He is cautious, making sure he is right before he speaks out. He is pragmatic, for the nature of the job demands that he be practical.

### Functions

The specific functions that the standards engineer is authorized to perform will vary depending on his place in the company's structure. The title he has will vary from one company to another. He may be the supervisor of, the chief engineer of, the manager of, the director of, or one of many other titles. His responsibilities are as diversified as his titles. Typical areas of responsibility are engineering standard parts, design techniques, drafting practices, engineering routines, installation procedures, specifications for components and parts, engineering policies and procedures, and so forth.

A recent survey conducted by the Washington Section of the Standards Engineers Society listed the various functions of the standards engineer. This survey indicated that his principle functions are as follows:

1. Develops and administers the standardization program

2. Prepares, coordinates, establishes, and maintains company standards

3. Assists in the development and maintenance of industrial standards

4. Serves as a consultant to other engineering divisions and as a corollary to this, administers the collection and maintenance of a library of standards, specifications, and other technical information

These are the main functions which emerge from a variety of positions described. The wording used varied widely. For example, regarding the standardization program, it was his function to establish, set up, initiate, supervise, develop, maintain, administer, participate, and interpret. The usage was even more varied in standards development, where the standards engineer might study, develop, establish, review, promulgate, assist, guide, approve, supervise, edit, revise, maintain, publish, coordinate, promote, write, recommend, formulate, investigate, compile, and propose.

This survey also revealed that the standards department or the standards man answered to the supervisor of engineering services, the application and evaluation manager, the chief engineer for industrial methods, the chief engineer for processes, the manager of standards and procedures, and the manager of technical services. The standards group is usually part of the engineering department, rather than being independent. Unfortunately, management support, if it exists, is through channels rather than direct. The latter method is often cited as an essential to a dynamic standardization program.

A summary of the documents for which the standards engineer is responsible was revealed as a secondary duty. The survey disclosed the following groupings: engineering standards and standard drawings for components and parts, design techniques, drafting practices, engineering routines, installation procedures, manufacturing processes, materials, products and test methods, specifications for components and parts, equipment, manufacturing processes, materials and test procedures, and design and drafting manuals. Long as it is, this listing does not include all of the documents which may be under his jurisdiction.

A number of features of the standards engineer's work were brought out in other descriptions. He serves as a consultant to engineering sections, to development and production sections, and to manufacturing sections. His function, he must keep abreast of new products and materials, procedures, processes, and systems. The third feature is that he must be able to meet and cooperate easily with people in other departments, manufacturing personnel, vendors, and with engineers from other companies. Although this might be expected of many employees in other employment areas, the emphasis in job description would indicate that these are particularly significant features of the standards engineer's work.

This then is the picture of the standards engineer and his duties as gleaned from the survey made by the SES Washington Section.

**Job Analysis and Evaluation**

Job descriptions begin with job analysis and evaluation – the complete process of determining the value of an individual job in relation to the other jobs in the organization. "Relative rather than absolute values of jobs are obtained."* Job descriptions involve setting a price on this relative value by establishing minimum and maximum salaries.

Job analysis is the process of determining the nature of a specific job by observation, study and tabulation of pertinent information relating to the job. It is the setting forth of the tasks, skills, knowledge, abilities, and responsibilities that the successful performance of the job entails and the features which differentiate the job from all others.

In order to give the word "job" precise meaning, two other terms must be defined: task and position.

"A task is created whenever human effort must be exerted for a specific purpose. The purpose may be physical, such as pulling and lifting, or mental, such as planning and explaining."* The effort may be exerted to change a thing or merely to maintain its quality. The thing may be as tangible as a hammer or as intangible as verbal expression.

"When enough tasks accumulate to justify the employment of a worker, a position has been created. A position, therefore, is an aggregation of duties, tasks, and responsibilities requiring the services of one individual."* A job may be considered as a group of positions which are sufficiently alike to justify their being covered by a single analysis.

When making a complete analysis of a job, the following four points must be considered: (1) What does the worker do? (2) How does he do it? (3) Why does he do it? (4) What skills are involved?

---

*Reprinted by permission from *Industrial Engineering Handbook,* by H. B. Maynard. Copyright, 1956, McGraw-Hill Book Company, Inc.

In determining what the worker does, you must establish the complete scope of the job and consider all of the physical and mental activities involved. How the work is done, means how the use of tools and how the movements of the worker himself are involved in accomplishing the task. The why of the job is the overall purpose. The skill factor refers to manual skill, knowledge, responsibility, mental application, dexterity, accuracy, experience, and training requirements. Skill is also considered to include the possession of certain unlearned qualities known as worker characteristics. Knowledge of these characteristics and their relative importance aids in the proper selection of workers.

As mentioned before, job evaluations and descriptions involve the establishing of minimum and maximum salaries for each group of jobs based on their relative value. The general wage level in the plant should be consistent with the going wage level in the community for comparable jobs. Also, the wage rate for each job or operation should be determined relative to other jobs in the plant, with due regard for differences in job content. It should be remembered that jobs are to be rated, not the employees performing the jobs. It is strictly an impersonal procedure.

Once the job has been properly analyzed it can be evaluated through the application of a measuring stick. The National Metal Trades Association plan consists of evaluating eleven factors. Each of the factors is divided into degrees and points are assigned to each degree in relation to the importance of each factor to the job as a whole. In other words, the factor values are weighted. The points assigned to the factors have been proved valid through years of experience.

1. Education – The basic knowledge or "scholastic content" (however it may have been acquired) essential as background or training preliminary to learning the job. The background may have been acquired by formal education, by outside study or by training on jobs of lesser degree. The rating is expressed in terms of equivalent formal education for convenience.
2. Experience – Experience measures the length of time usually or typically required by an individual, with the specified educational background, to learn to perform the duties effectively under normal supervision. In rating a job on this factor, it should be remembered that experience is of two kinds.
    a. Previous qualifying experience on related work or lesser jobs, either within the organization or outside.
    b. The "breaking-in time" or period of adjustment and adaptation on the specific job itself.

Both periods must be added together to secure the over-all rating on experience.

3. Complexity of Duties – Measure the complexity of the duties involved, the degree of independent action, the extent to which the duties are circumscribed by standard practice, the exercise of judgment and the type of decisions made, the amount of resourcefulness and planning the job requires, the creative effort in devising new methods, policies, procedures or products, scientific discoveries and original applications.
4. Supervision Received – The degree to which the immediate superior outlines the methods to be followed or the results to be attained, checks the progress of work or handles exceptional cases. Consider the proximity, extent and closeness of supervision in rating this factor.
5. Errors – Measure the responsibility for preventing errors due to carelessness. Consider the probable effect of errors based on the degree to which the work is verified or checked, either in succeeding operations, by the procedures themselves or by supervision. Consider the probable monetary loss, damage to equipment, labor and material costs for correction, customer loss or good-will, etc.
6. Contacts with Others – The responsibility which goes with the job, for meeting, dealing with or influencing other persons. In rating this factor, consider how the contacts are made, how often, whether contacts involve furnishing or obtaining information only, or whether they involve influencing others.
7. Confidential Data – Measure the integrity and discretion required in safeguarding confidential data handled or obtained in the normal performance of duties. In rating the job, consider the character of the data, the degree to which the full import of the data is apparent on the job in question, whether disclosure would affect internal relationships only, or external, competitive relationships.
8. Mental or Visual Demand – Measure the degree of concentration and co-ordination of mind and eye. Consider the volume or flow of work, the requirement for co-ordinating manual dexterity with mental and/or visual attention and the sustained character of the visual attention. This factor should not be used to measure the qualitative or quantitative requirements of the job for "headwork."
9. Working Conditions – The surroundings or physical conditions under which the job must be done and the extent to which they make the job disagreeable. It will have little application except on sales, outside service, factory supervisory jobs or

other occupations where duties are performed wholly or partially in the shop.*

Factors 10 and 11 are not defined since they pertain to supervisory personnel.

### Grading the Jobs

The measuring stick is then applied and the job requirements for each factor written up and the degree and point values inserted. The write-ups must substantiate the degrees assigned and are usually referred to as "substantiating data." The point values for each factor are totaled and this sum constitutes the total point score of the job. The job is then placed into a grade which is determined by the total point score. Each job should be measured and rated in accordance with the same procedure and placed in grades. At the completion of the survey all jobs falling in the same grade are comparable jobs and should be subject to the same pay scale or rate.

Having reviewed briefly some of the characteristics and qualities of the standards engineer and having touched on job evaluation and descriptions in general, it is now possible to begin preparing a job description for a standards engineer. Some basic rules for style to be considered when writing job descriptions are:

1. A terse direct style should be employed.
2. Each sentence in the "duties" section should begin with a functional verb.
3. The present tense should be employed throughout.
4. All words that do not impart necessary information should be omitted.

### Job Description – Standards Engineer

*I. Education*

A. BS degree in engineering or equivalent experience

B. Thorough familiarity with principles of standardization

---

*C. C. Blumer, *Job Evaluation,* National Metal Trades Association, Chicago, Illinois.

C. General knowledge of engineering methods, procedures and general administration
D. General knowledge of design, drafting, and development practices
E. General knowledge of manufacturing, materials, parts and processes
F. Thorough knowledge of specifications, recording and nomenclature
G. General knowledge of inventory control, accounting systems and inspection methods

*II. Experience* – Minimum of four (4) years experience in related engineering field. Preferably within your own company.

*III. Desirable Personal Attributes and Skills*

A. Ability to organize, plan and control work and perform expeditiously and efficiently
B. Good analytical mind
C. Initiative and foresight
D. Ingenuity and good judgment
E. Enthusiasm and good imagination
F. Tactful, yet forceful
G. Ability to make decisions
H. Resourcefulness
I. Personality adapted to oral and written self-expression

*IV. Nature of Duties*

A. Initiate investigations of subjects or areas where standardization would be beneficial to the employer
B. Gather material and technical information for study, analysis and the eventual formulation of standards and specifications for consideration by the standards committee
C. Act as chairman of standards committee or subcommittee as directed
D. Prepare agenda for standards committee meetings and provide the required follow-up, editing, etc., necessary for the approval and publishing of a standard

E. Prepare, publish, distribute and maintain standards for the approval and publishing of a standard
F. Analyze and offer solutions to problems in his specialty field
G. Interpret the significance of standards of specialty field
H. Perform liaison with various operating departments of the company, keeping them informed of latest materials, techniques, processes, etc., regarding standards
I. Maintain indices and files of commercial catalogs, standards manuals, society standards, specifications for parts and materials, and other relevant technical data
J. Check for proper use of standards within the company and report any irregularities in the program
K. Recommend improvements to the program
L. Act as consultant on standards to all levels of management to further the standards program
M. Promote, on a supervisory level, standards and standardization within the company
N. Conduct test and evaluation programs to determine design data suitability and applicability of new standard parts
O. Travel periodically to participate in standards committee work on a national or industry level
P. Keep abreast of latest engineering and commercial development through book and magazine research, national and industry standards, and exchange of information with other companies

*V. Supervision Received*

A. Under direction of the Chief Standards Engineer, working from company policies and general objectives of the standards department
B. Must be capable of operating with a minimum of supervision. Usually only contacts superior when clarification or interpretation of company policy is involved

*VI. Errors*

A. Possible errors may involve major expenditures for equipment, material or products or loss of important customer accounts

B. Duties may involve the preparation of data on which top management bases important decisions; erroneous data could be very costly

*VII. Contacts*

A. Works with people within the company as well as personnel from other companies. Requires a high degree of tact, judgment, and the ability to deal with and influence persons in all levels

*VIII. Confidential Data*

A. Regularly works with confidential data of major importance, which, if disclosed, could be detrimental to the company's interest

*IX. Mental & Visual Demand*

A. Flow of work and character of duties involve visual attention and occasionally unusual concentration
B. No unusual physical demand

*X. Working Conditions*

A. Normal office conditions. Seldom exposed to disagreeable elements. Accident and health hazards negligible

The job description shown above is quite comprehensive. However, it is entirely possible that some of you might wish to add to the information to satisfy the special needs of your company or method of operation. It has been prepared in such a manner that adding to it or subracting from it can be done with a minimum of effort.

The establishment of sound job descriptions is an important step toward a general recognition of the importance of the Standards Engineer in the corporate picture. It is realized that this brief article couldn't possibly bring this recognition about. However, it does give you some guide posts that will be helpful to you in preparing job descriptions of your own. You may be assured that the time and energy spent will not be wasted.

## References

*Industrial Engineering Handbook, First Edition*. McGraw Hill Book Company, Inc., New York, N.Y.

National Metal Trades Association

*Job Descriptions*, Studies in Personnel Policy, No. 72, National Industrial Conference Board, New York, N.Y.

*Standards Engineering* magazines. Published by Standards Engineers Society

"Shirtsleeve Standards" by A. M. Rosewald, in *Standards Engineering*, published by Standards Engineers Society

My thanks to the many standards engineers and companies that expressed interest in the subject and submitted information for consideration.

# ABSTRACT OF REPORT ON EDUCATIONAL COURSES FOR STANDARDS ENGINEERS

Rowen Glie

The Board of Governors and the Educational Committee of the New York Section of the Standards Engineers Society decided to call a conference of the managers and supervisors of standards departments to define the academic requirements for the function of a standards engineer.

The importance of the subject cannot be overemphasized. Not only we, the managers of standards departments and sections, have great difficulty in finding engineering talents with the required qualifications, but should an engineering student decide to specialize in standardization there is no institution that would truly provide it. Should some institution decide to create a curriculum for the title of standards engineer, it would be impossible for them to do so because they would not know what is required in order to function as a standards engineer.

The answer to the whole situation obviously lies with the managers and supervisors of standards engineering departments and sections. Only the ones who hire engineers, and make standards engineers out of them, manage them, make them do the desired work, rate them and promote or fire them, only they know what kind of knowledge is necessary and can determine what should be the academic requirements for a standards engineer.

Creation of a curriculum necessary for a standards engineer will also have a great effect on the job description and salary qualification as well as on the status of the standards engineer within the engineering organization. It will start a chain reaction, the result of which will be that we will understand ourselves better, our functions will become defined, our position in the company stronger, and, last but not least, our salary will be on a higher level.

At the present time, it is too often that a standards engineer is a self-made man and, therefore, does not have well-defined or well-understood functions, and is looked upon as some kind of an engineering vegetable, not meat nor fish.

Before an educational program for standards engineers can be formulated some premises related to what a standards engineer is and what he does must be accepted.

1. My personal premise of what a standards engineer is, is as follows: A standards engineer is a staff man in charge of repeatable tasks in the company, such as in engineering, manufacturing, design, drafting, and, in general, in all repeatable activities of the company.

2. Considering the whole company as a system whose purpose is satisfaction of company's type of customers, the standards engineer examines all repeatable activities of the company necessary to complete the tasks and determine on a company-wide basis the quantity, type, and quality of materials, components, processes, etc., which would be satisfactory to both the company, from economical point of view, and company's customers, from service point of view.

3. My third premise is that a standards engineer must be a graduate engineer or a physicist, or have equivalent experience and, therefore, a full educational program for standards engineers will obviously include subjects which are already covered by his previous education. For instance, probability is already included in the curriculum of most engineering schools, and, therefore, would not be necessary as a special course unless it was not included in the curriculum of a particular school.

4. The educational requirements for components and standards engineers are the same on so many different points and different on so few, that the same program should suffice for both types of engineering and the differentiation will start on the job itself.

A curriculum of standards engineer actually presents a list of subjects of which only certain additional credits will be necessary for each particular graduate engineer in order to obtain the title of standards engineer.

Attached is, for your perusal and as a basis for further discussion, a tentative program of courses of standardization which are necessary for an engineer specializing in standardization.

**Program of Educational Courses for Standards Engineers**

*I. Company Management Fundamentals*

A. Business organization and economics, which should include
   1. Fundamental business organization
   2. Planning
   3. Functional control
   4. Company controls
   5. Organization structure, etc.

B. Business economics
   1. Concept of benefit and cost
   2. Budgeting
   3. Price policies
   4. Forecasting
   5. Statistics and probability as a tool of management

C. Contract administration
   1. Laws of military and nonmilitary procurement
   2. Contract clauses
   3. Proposals
   4. Price determination
   5. Subcontracts
   6. Compliance with customers' (government and non-government) standards

D. Factory and administrative systems and procedures, including stock control and inventory management

E. PERT/CPM

*II. Company Standards and Components Engineering*

A. General introduction which shall include
   1. Definition of standardization and specification, area of application, short universal history of standardization (Egyptian cubits, Mass production Venetian galleys)
   2. Standards and modern means of production requiring interchangeability and predictability (reliability): quality and mass production, automation, numerical control
   3. Standards and commerce – trade standards, importance of standards to import and export, international standardization

4. Standards and communication – scientific technique and engineering symbols, terminology, abbreviations, drafting standards
5. Standards and quality assurance, testing standards, and specifications
6. Standards and medicine – American Pharmacopoeia
7. Standards and science – Measurement standards, activity and services of the National Bureau of Standards
8. Material standards and process specifications
9. Main American standardizing agencies – USASA, military, federal, Electronic Industry Association, NEMA, ASME, IEEE, SAE, AISI, ASTM, International Standards Organization (ISO)

B. Study of indices and standards formats and contents of the typical standards issued by the above mentioned organizations

C. Company standards – Their necessity as a means of
1. Management control of company repetitive operations (engineering, manufacturing, purchasing, etc.) by recording and reuse of company know-how; advantages of standards are apparent in engineering, manufacturing, purchasing, testing, drafting, stockrooms, etc.
2. Company components engineering for optimization of the quality and reliability levels combined with minimization of cost
3. Company-wide utilization of company and state of art experience with components

D. Techniques of standardization
1. Fundamentals of determining labor cost-standard time concept
2. Comparative cost of different materials and processes used by the company
3. Cost of close tolerances and of high reliability and confidence levels
4. Stock room administration and inventory control
5. Fundamentals of procurement
6. Standardization methods-simplification, value analysis
7. Formulation of standardization findings in standards and specifications; layout, format, contents, revisions, etc.

8. Enforcement of standards policies
9. Appraisal of job contract requirement as related to standards, and specifications, materials, drafting, etc.
10. Study of major government and nongovernmental standards and specifications

E. Techniques of components engineering
1. Specification control drawing – purpose and preparation – different types as required by military specifications
2. Evaluations of vendors – methods of vendors' ratings formulas
3. Fundamentals of components testing, including study of available testing standards and specifications
4. Appraisal of vendors' products

F. Technical writing
1. Principles of expressing ideas clearly
2. Reports, standards, and specification writing
3. Page sizes, page layout, illustrations, bindings, covers
4. Printing methods (multigraph, mimeograph, offset, Xerox, etc.)
5. Distribution and revisions

G. Organization of a standards department and components department
1. Different functions and performance requirements in the standards department
2. Job description for various positions from technical help to manager
3. Duties of standards engineers, job description, etc., as above
4. Relation of standards to other departments – reliability, purchasing, design, shop, etc.
5. Organizational charts for standards departments
6. Preparation of a yearly standards department program
7. Preparation of standards department budget
8. Progress (status) reports

H. Sources of engineering information for a standards and components department
1. Library of military, federal, and other specifications and vendors' catalogs, technical data, failure reports, reports from IDEP, PRINCE, etc.

2. Military specifications, NASA and commercial requirements for components inspection and testing (knowledge of available military and technical test specifications)
3. Fundamentals of operation with EDP equipment
4. Use of computers to solve problems of standards and components engineering

*III. General Engineering Studies*

A. Mathematical refresher – calculus, matrices, symbolic logic, etc.
B. Fundamentals of probability and statistics applicable to reliability, testing, quality control
C. Metrology and dimensional analysis
D. Theory of models
E. Reliability, maintainability, DC, inspection – organization, functions, responsibility, sampling, acceptable control levels, failure reports, reliability assessment, etc.
F. Technology of electrical and mechanical components
G. Physics of materials (fundamentals of metallurgy and plastics) and effect of environment on materials
H. Fundamentals of servo mechanisms, feedback concept, (open and close loops), automatic controls
I. Mechanism failure – why and how components and materials fail, including failure reports and analysis as well as probable corrections
J. Electrical, electronic, and hydraulic circuits (fundamentals of)
K. Analog and digital computers, application to design, drafting, standards, and components engineering and solutions of engineering problems

# TRAINING OF ENGINEERING AND TECHNICAL PERSONNEL IN A STANDARDS DEPARTMENT

John Rayfield

In order to understand and appreciate my talk on "How to Train a Standards Engineer," I would like to establish my parameters of a standards section's responsibilities. Depending on the company, standards work is carried on under many different titles, such as standards engineer, components engineer, applications engineer, quality engineer, reliability engineer, chief draftsman, chief designer, and many others too numerous to mention. This being the case, it is obvious that we must tailor the standards section to suit the company.

For an effective standards department the recommended minimum areas of interest and responsibilities are as follows:

1. Manuals: drafting, design, parts, materials, and process
2. Specifications: military, industry, commercial, and company
3. Standards: military, industry, vendors, and company
4. Components: military and commercial
5. Materials: preferred and stock sizes, warehouse and mill sizes
6. Processes: manufacturing procedures
7. Test & evaluation: components, materials and processes
8. Quality & reliability: components, materials and processes
9. Nonstandard parts lists
10. Consultants: proposals, contracts, purchasing, quality control, reliability, manufacturing, design, drafting, and engineering
11. Standards library: military specs, QPL's, handbooks, catalogs, process specs, and vendor prints
12. Educational programs: seminars and demonstrations

On the basis of the foregoing, the following is a recommended procedure for training of personnel in standards department routine. For the sake of preparing a complete system of training we will consider the trainee as a beginner; that is, one who has had little or no experience. The selection of a person who is suitable for training is of paramount importance and should be carefully analyzed in order to make the most suitable choice. The following requirements should be considered in making the selection:

1. Education: Engineering degree preferred or technical junior college; or two years college education in professional engineering and a few years' experience as indicated below; the ability to write good English is essential
2. Experience: Draftsman, designer, laboratory technician, inspection, or equivalent technical experience
3. Personal: Clean, neat, nice dresser, good speaking voice
4. General: Interested in learning and advancing his knowledge, likes to read, mechanically inclined, must have initiative and imagination

Having suitably screened and selected a prospective standards trainee, the training of the selectee is most important for the proper end result.

To start the training he should be assigned to the standards library. The standards library contains all the catalogs; military and company specifications; military, industry, and company standards; process specifications; and military and commerical indices, and is an excellent place to become familiar with the various documents used in standards work. Under the direction of the librarian or supervisor he is taught what is in the library, where the various documents are filed, and how to find and charge out the document. He is further taught how to give service in obtaining a document such as catalogs, specifications, standards, etc, from their respective sources. In order to give this service he must become familiar with the various indices such as DOD, Army, Navy, Air Force, ASTM, AMS, NAS, NAF, Conover Mast, McRaes, Buyers Guide, etc. These indices are the means by which these documents can be identified as to their source and where to obtain them. In order to become familiar with components materials, equipment, etc., it is important that, after becoming familiar with the various catalogs, he read the various magazines and periodicals delivered to the standards section, and, as required, send for catalogs of new items as well as new catalogs to update those in file. The Standards Library is normally responsible for keeping, borrowing, and returning catalogs, specifications, and standards, and recalling and replacing obsolete or revised specifications, standards, and catalogs.

Having mastered these preliminary tasks in the Standards Library we are now ready to put this basic knowledge to work as outlined in the paragraphs that follow.

Show him how to use the copying machine for copying documents, letters, drawings, technical data, etc. In particular, he should be taught how to make a direct hard copy, paste up hard copy, and in general make drawings on reproducible vellums on the copy machine without actually drawing.

Show him how to use the Military Indices to look up an item by subject or number; how to find a component, material, equipment, standard, QPL, custodial agency, and where to write for the document.

Become familiar with the various agencies and where to obtain the various military documents. It is also advantageous to become familiar with specific people responsible for the various activities of the agencies. For example, Mr. John Jones in charge of connectors, Mr. Sam Huston for adhesives, Mrs. Clara Simmons for specifications, Miss Lora O'Brien for test data.

Have him prepare a simple company standard of a part for one of the standards engineers. Before preparing the standard, show him how to go about looking for suitable vendors making this part. See that he calls or writes to these vendors for their catalog, technical data, specification, material, part number, whether the items are in stock, delivery schedule, and other pertinent data. Proceed to show him your standard format for preparing such a drawing; what to draw or, preferably, copy from the catalog using the copying machine, where to place the parts, tables, purchasing and inspection requirements, notes, approved vendors' names and parts numbers and any other required data. Show him how to select and assign your company part number. He should be shown how to select titles from military document H6-1, and vendors code identification number from H4-1k. Now that the standard document has been prepared, the trainee should be taught to see that the necessary approvals are obtained and finally released through the established company release system.

In connection with the preparation of the standards parts drawing he should be taught how to select approved vendors by any one of the various accepted methods such as physically testing the part; obtaining acceptable test data from one of the military or private agencies; contacting an associate company who may have already tested the item. He should also be made to understand that there are existing documents for testing such as MIL-STD 002, ASTM and many others.

He should be taught what an advance material release (AME) is and the responsibility the standards section has in checking and releasing them. Have him check simple items at first and progressively difficult ones as time goes on. See that in checking the items listed that knowledgeable items are quickly dispensed with while questionable items are thoroughly investigated to see that parts numbers are correct, military and/or industry specifications are noted; that standard items are recommended where nonstandards are specified.

Starting with simple problems and slowly progressing to more difficult problems which arise from day to day in various parts of the company for which the standards section is responsible, the trainee should be assigned to overcome these problems. For example, the purchasing department has ordered some paint to a military specification. When received there was no certification, there is no marking or identification for conformance to the specification, and the inspection department desires to know if it is military material. If it is, how do they accept it. If it is not, how do they reject it. The trainee should be taught how to investigate, test, clarify, and certify if it is or is not acceptable material. He should also be taught the proper steps or procedures to accomplish this.

Another very important phase for the trainee to know is how to prepare a specification control drawing. Before he can prepare this drawing, he must be made to understand that a part must meet certain requirements in order to be considered a specification control part. You cannot arbitrarily class all parts as specification control parts. You must be taught how to recognize and differentiate a specification control part from other parts. He should know that a specification control part is one that is a commercial part of a vendor's design, normally stocked or manufactured by the vendor as his stock or catalog item for which the vendor has assigned a part number. The specification control drawing shall indicate only that data with which the vendor's drawing, catalog, specification, describes his part. After he has been thoroughly indoctrinated in the area of specification control drawings, he should be assigned for a period of time.

Learning when and how to contact vendors, sales representatives, military and government employed personnel, colleges and universities, technical societies and organizations, magazines, etc., are of paramount importance to know. If we would take the case of preparing a simple specification control drawing, he would want to know who and how many different vendors are making the part; available materials; mechanical and electrical capabilities and characteristics; finish; is the mounting hardware delivered with and/or part of the component or must it be ordered separately; is it readily and/or locally available; does it conform to military requirements;

does the vendor have and will he submit substantiating test and evaluation data; does the part have built-in quality and reliability and ever so many other pieces of information. To obtain all this data he would have to investigate the part to see who makes it; contact the vendor (s) for all the available data; if necessary, test the part to the degree required. If there is conflict between the vendor and the military requirements, wherein misunderstanding or interpretation of the requirements is confusing, they must be clarified. To clarify these requirements he must contact the responsible agency and in particular speak to the person who is responsible for the document and, if possible, preferably the person who wrote it, and inquire the true intent of the area in question. When the data is verbally obtained, have them document the answer in writing for future reference and confirmation. It is normal that, in order to have your answers documented, you will have to write the agency requesting confirmation of the verbal discussion discussed with Mr. XYZ on the particular subject.

To assist draftsmen in the proper preparation and execution of an engineering drawing the standards section issues a drafting room manual. This manual contains the necessary information on how to draw or make a drawing. This manual should be a primer for a trainee. He should become so familiar with this manual that he knows it backwards and forwards as most of the trouble in drafting departments is that they do not know how to make a drawing, or, to say the least, not in compliance with the standard method and system outlined in the manual. Having a schedule and commitment to meet, group leaders assign work to draftsmen; they try to do the job as fast as possible and, therefore, could not possibly complete a drawing and do it right within a short time without taking short cuts. The trainee should understand that the standards section prepared the drafting room manual and is constantly called upon for clarification, counseling, revision, etc., and therefore, without knowing this manual, would be in poor position to defend it. Therefore, the standards engineer becomes the expert and/or consultant and he should know his drafting room manual.

To assist the designers and engineers, the standards section issues a design manual. This manual contains information on how to design. The trainee should become familiar with this document as to its purpose, contents, and future contributions of additional data that he will be required to make, be it revised or new. Towards this end he should be indoctrinated in the manual by assigning him some simple task in preparing design data such as machined centers or knurling. Have him prepare a design standard by giving him parameters or guide lines as to what is the purpose of the data; what areas of interest should be included; how do you communicate this data to the engineering drawing; how do you specify same so that the manufacturing, inspection, quality and reliability departments know exactly what is required; where to obtain the data; and many others too numerous to mention.

Information contained on the engineering drawing must be documented in such manner that perfect communication to the other departments is clear and understandable. In particular the manufacturing department is concerned with the manner in which processes are specified. Processes are actually specifications which spell out how to accomplish a specific operation or a series of operations. To avoid lengthy explanations and requirements which would be required to be placed on the engineering drawing, process specifications are created. Each process specification is coded so that a simple code such as number 101 placed on the engineering drawing would invoke a mountain of paperwork containing data which normally would take time to print on a drawing. The standards section is responsible for preparing, disseminating, and revising these process specifications. The trainee should be made aware of the responsibility and importance placed on these process specifications. Let us start the trainee off with a simple process specification which, for the sake of illustration, we will code 101 and the process is "vapor degrease." The specification would include a scope, purpose, material, military spec., if any, solution, method of mixing, operating temperature, precautions, procedure, application, and inspection. To the manufacturing department, when 101 is specified on the engineering drawing it infers cleaning by vapor degrease. The trainee should be made aware that to prepare a process specification requires extensive investigation, reading, testing, coordination with other departments, equipment availability, cost savings between doing it in plant or subcontracting. It should also be remembered that corrections or changes to these process specifications are constantly in process and form a very important part of updating the process specifications.

A very important function of the standards section is preparation of the nonstandard parts list. In all military contracts this requirement is invoked and usually is confined to electronic components. The trainee should be made to understand the difference between a standard and nonstandard part, particularly since many military contracts have as many as 60% nonstandard parts. By understanding that any military part such as an MS, AN,

JAN, etc., is a standard part and that all others are nonstandard parts, it makes it easy for the trainee to differentiate. Depending on the contract, he should be taught that there is a typical example of a nonstandard part approval form to be used and that it must be suitably prepared and correctly filled out so that all pertinent data required appears on the form. The trainee should be made to understand that these nonstandard parts must be listed individually on separate sheets with necessary substantiating documentation attached. When properly indoctrinated, the trainee should be capable of taking an assembly drawing and, by looking at the bill of material, select all the nonstandard parts, prepare the necessary lists, have them certified by the responsible engineer, obtain the necessary pertinent technical data and attach it to each of the lists and, through the contracts department, submit them to the cognizant military agency for approval.

One of the most important functions of a standards engineer is selection and application of components and materials. The trainee should be indoctrinated in easy, simple stages, starting with components like screws, nuts, and bolts, and slowly progressing to more difficult components, including raw materials. The knowledge acquired from catalogs on components and materials when training in the Standards Library will be of tremendous value in selecting and recommending components and materials. This phase of the trainee's education should be controlled by the standards section head, who will assign him to each of the standards engineers to assist them in various fields and applications, thereby gaining valuable experience. As a result of this experience the trainee will become proficient in the standard and nonstandard parts – in particular, the preferred parts and materials.

In conjunction with selection and application of components and materials is testing and evaluation. It is of the utmost importance that components and materials which are preferred and recommended for use in military equipment are reliable. The trainee should be assigned simple testing projects which are progressively increased in scope and responsibility. Eventually this phase of training will be of tremendous value when recommending reliable components and materials.

The time required to implement and accomplish this training program depends on many factors, i.e., organization, manpower, facilities, responsibilities, and on the man who takes the training. However, I believe that a minimum training program can be accomplished in 4 to 6 months. What must be emphasized is this – whatever the efforts and time necessary for training are, a training program is essential.

In summing up my talk on training in standards it should be remembered that self-education is a very important part of training as well as of continuous growth in professional competence of the technical member of the standards department. The trainee should be encouraged to attend night college and work towards a degree, or take specific courses in his particular field of interest. Likewise it is important that he be encouraged to join various technical societies and actively participate, particularly in Standards Engineers Society. As of now, there is no degree for standards engineers, and there is no place where you can hire a trained standards engineer, or where a young interested beginner can go to school to be trained. It is, therefore, up to each and everyone of us to realize that we have to institute a training program in each of our firms until such time as there will be schools to teach and/or train a standards engineer and, therefore, from here on it is up to you.

# IS THERE A DIFFERENCE?

*(The Role of the "Engineer in Standards" and the "Standards Engineer" in a Medium-size Decentralized and Diversified Industry)*

Fulton R. Magill

Strange, isn't it, that there should be any questions about the position of the "engineer in standards?" Standardization may not have been born in engineering, but it has certainly lived with it for a long time; and engineering has exerted a paternal guidance and control in standardization work. For this reason, it is important that we understand the distinction between the "engineer in standards" and the "standards engineer."

In general it might be said that the engineer in standards is that engineer who actively participates in selecting those technical or design details which are to constitute a national or company standard.

The standards engineer, on the other hand, is a man concerned with the administration and coordination of this standards work in a manner prescribed by management.

In short, the standards engineer deals with people and data, whereas the engineer in standards deals with specialized engineering or scientific knowledge.

To prevent confusion, before continuing this article, it is necessary to emphasize its limits. It considers only a decentralized industry with diversified products. The industry is medium size, having annual gross sales of more than 50 million dollars.

We are concerned only with portraying the differences between the functions of the engineer and the standards engineer in standardizing work in this type of industry.

The term "engineer" is being applied to a member of a learned scientific or technical profession who by his training and experience is qualified to participate in establishing the technical details required by a standard.

The term "engineer in standards" is applied to the engineer who actually participates in the formation of standards. The extent of his participation is mentioned later.

The word "standard" refers to any written answer to any problem, obtained by a consensus.

Changing times have affected the engineer's role in standardization. With expanding industries there has been decentralization of operations and a diversification of products. Standardization no longer affects only drafting and design practices within a small, integrated plant. The old rules established by specialists in one plant may not be acceptable or practical in decentralized and diversified operations.

Engineering has become more specialized while company standardization work has become more comprehensive. Generally, the standardizing efforts encompass the interrelated functions of engineering, manufacturing, and purchasing in different divisions. To understand the position of the engineer in standards, consider the scope of the standards program.

In a medium-sized decentralized company which produces a diversified line of products, there may be hundreds of different products manufactured; however, they probably will fall within a half dozen basic classifications. These products will be manufactured in from 2 to 30 different plants scattered throughout the country. Product components may be made in one plant and used in another. Standards are needed to allow these operations to work together.

The standardizing effort will be designed to reduce costs through the creation of a set of standard materials; engineering design standards, drafting standards, standard material specifications; standard parts; standard ingredients, tools, etc. To accomplish this standardization, a department of from two to eight people will be needed.

This department will collect data and facts from practically all departments and divisions because its standards are "engineering-purchasing-manufacturing" standards which affect them all. These data must be correlated. Data and facts from outside the company must be correlated with the internal standards work. The results are published in books of company standards.

In any company standardization work, it sooner or later will be necessary to justify the cost. The savings must be greater than the cost.

Management will be mostly interested in the standardization of product parts. Here is where management can find real savings.

The engineer wants design and drafting standards. If the parts are to be made in different divisions – if they are to be interchangeable – there must be design standards. If the drawings are to be made in one division and read in another, they must be as easily read in one as the other. In order to have the standard parts with their proven savings, we need the design and drafting standards.

We have mentioned reading drawings and making parts. This takes place outside the engineering departments and actually means that the engineering standards are, in effect, manufacturing and purchasing standards. The men who read the drawings influence the standards which govern the making of the drawings as much as those who do the drafting. There are very few standards which apply only to engineers.

The engineer has an important role in standardization, but he has another definite responsibility within his company which is of prime importance. He must design products or tools which will maintain his company's position in a competitive field and assure both management and the customer a reasonable return on each dollar invested.

He performs his work within an engineering organization which is usually of a straight-line military type. Working within the limits of his responsibilities and organization, therefore, can the engineer also participate in a function which not only cuts through all other line organizations but occasionally extends beyond company limits?

By so doing, it might seem that the engineer is exceeding his proper function – that is, working outside his organization to contribute to the standardizing effort. It might seem he cannot do this and still perform his normal duties. But this is not true. He does not work outside the boundaries of his organization. He leaves that for the standards engineer.

The standards engineer is a member of a newly recognized profession. Any medium-size company will find it practical to use him. However, the standards engineer will have neither the time nor talent to create, by himself, all the standards which are needed. He will concentrate on obtaining answers and analyzing them. He will coordinate the work – with the full understanding that standards affect sales, advertising, accounting, industrial engineering, maintenance, manufacturing, and purchasing, as well as engineering. He will occasionally go outside his own company to find some of the material for company standards.

The standards engineer knows that standards which have value – standards which will be used – must represent the thinking of the persons who use them. He knows that in his work no one department – no single person – has more influence than another. If a problem affects only engineering, it will be handled by engineering. The same applies for manufacturing and purchasing. The standards engineer is simply the administrator of a planned effort to coordinate answers to problems which affect all three functions.

The design or plant engineer plays his important part in company standardization when he furnishes the original technical information needed to create a standard. He also makes suggestions for changes needed to assure approval by manufacturing and purchasing. And he still functions as an engineer within his organization.

As an engineer in standards, he has still another important role. Every company is influenced by national and international standards. These standards are technical or engineering standards of various kinds, and it is impossible for the standards engineer as such to participate in forming them. The qualified company engineer is in a position to analyze the effect of certain national standards work on his company; and in some instances he should actually participate in formation of national standards.

This participation in national standardization by the engineers of a company must be controlled. In a diversified company there will be several engineers devoting part of their time to serving on different standardizing committees of national importance. Each engineer is a specialist in some class of engineering or product design. Each standardizing committee specializes on certain types of standards.

Since national standards are interrelated – and since a company's standardization program covers the interrelationship of products, divisions, and certain departments – the engineer in standards who represents his company's interests on a national committee must know how his committee work affects his company as a whole. His activities in his company are restricted to a specialized phase of engineering. Some agency must exist to keep him informed of the impact of his standards work on other things which normally do not concern him. The standards department, or the standards engineer who is the administrator of this part of a company standardization program, is responsible for keeping him informed.

With well-balanced participation in national standards work a company can be sure these standards will be practical, beneficial, and definitely not detrimental. The engineers in standards who serve in this matter may be considered part of the "State Department" of a company standardization program. This added function of theirs is vital – and adds an exciting and important new facet to the profession of engineering.

# Part 3

# STANDARDIZATION PROGRAM

*Why a company should have a standards department can be seen from different standardization programs presented here. After perusal it will become obvious that a standardization program is actually a program of management of company routine operations. It is only when management decides that a company can exist and prosper without management of its operations that a standardization program and, therefore, a standards department are not necessary.*

*R.G.*

# SELLING STANDARDS TO MANAGEMENT

*What Management Should Expect from a Sound Standardization Program*

Irving O. Miner

You are, I am sure, primarily interested in the benefits a standardization program can provide for your company. While individual company answers cannot, of course, be provided in this article, our experiences at B-I-F may serve to illuminate some possibilities and probabilities for you. I will try to be as specific as possible – no glittering generalities.

A little background on B-I-F may help you translate our experiences to your problems. Our company was established in 1820 as the High Street Furnace Company, its principal products being cast iron columns for textile mills and cast iron store fronts. In 1853 it was incorporated as Builders Iron Foundry.

The character of the company did not change much until around the turn of the century. In the early nineties Clemens Herschel started the development of the venturi meter for the measurement of large flows of water at Holyoke, Mass. He had his first castings made at Builders Iron Foundry . . . and someone at Builders had sufficient foresight to acquire rights to Mr. Herschel's invention. Today we find that the venturi is used by nearly all large municipalities for metering public water supplies.

The next problem was chemical. A city such as Providence has had to add many tons of chemicals annually to the public water supply to process it properly. Most of these chemicals should be added in proportion to flow.

A couple of young, capable, and vigorous fellows in Chicago were struggling to establish a chemical pump business during the early days of the depression. Since these pumps could be paced by venturi tubes to feed chemicals in proportion to flow, Builders Iron Foundry arranged to move this new pump company, Proportioneers, to Providence and finance its growth. Thousands of Proportioneers pumps were sold to municipalities, and even more of them were sold for industrial use. However, because the bulk of waterworks chemicals are fed in the dry state, such equipment was sought for the third "leg of the stool."

The leading manufacturer of dry chemicals feeding equipment was the Omega Machine Company in Kansas City. During World War II, the excess profits tax on top of the almost confiscatory income tax that is ever with us made it well-nigh impossible for a highly successful company to finance its growth. As a result, Builders Iron Foundry was able to buy the Omega Machine Company. We acquired all drawings, patterns, jigs, fixtures, and some machine tools. We also imported the principal executives and administrators.

Another big change came in 1955. The city had gradually grown around the foundry, making relocation essential for efficient operation. However, the foundry was now a minor part of our operation and was becoming steadily less important. Therefore, in 1955, instead of relocating, we divested ourselves of our foundry. A name change seemed in order; we adopted B-I-F industries.

In 1957 we found ourselves in enough of a cost-price squeeze to look for new ways to increase efficiency. Proportioneers was then wholly owned, as was Omega from its first acquisition. Different groups of people were still developing and designing the products of the different divisions, and it was evident that standardization offered a golden opportunity for savings. In the following paragraphs I have outlined our goals and expectations first, then followed with data on the actual realization.

**Time Saved by Designers and Draftsmen**

When selecting an item such as a ball bearing, a man had to dig out one or more catalogs and look for a suitable bearing. If he didn't like some feature, such as the type of seal offered by the first one or two manufacturers, he would get more catalogs and keep looking. When he figured the load rating from manufacturer A's catalog, it was done one way. In manufacturer B's catalog, it was done a different way. By providing a standards book with bearings that would fit over 90% of our requirements and with a single method of figuring load ratings, we expected to eliminate looking for catalogs, looking through more than one source of information, learning more than one way to figure load ratings, and hunting for a type of seal fitting a designer's particular whim. By putting the frequently used items in one book we could eliminate going to a central file containing many feet of catalogs. (No one is supposed to take anything from our central catalog file without signing for it, but if any reader knows of a central catalog file without a full-time librarian, where missing catalogs can always be located immediately, I would like to learn his secret.)

*Accomplishment*

We make sure there are always enough standards books so that if a person's own book cannot be found immediately, another is practically within reach. While I have no direct way of measuring the savings, I do know that, with about 25% fewer designers and draftsmen, we are handling the same dollar volume of orders and a greater number of orders than we were two or three years ago. Part of the reason is surely the standards book, although I think that setting up some of our standard products so they could bypass the design and drafting dept. is probably a longer cause.

**Reduction of Inventory**

Without a standardization program, designers and draftsmen will seldom select the same purchased item for a given need. Alert engineers are always on the lookout for improvements that they can incorporate into a product. Some look mainly for quality, others for economy. Thus a new product is apt to offer ever a new design of pressure-reducing valve or gauge or oilless bearing, while the drawings and bills of materials for previous products still call for the older parts. The production control department must stock a large variety of such items to meet delivery schedules. With a standards book carrying properly selected items offering an optimum balance between cost, quality, and features, a substantial reduction in item variety and inventory was looked for. It was expected that the standard parts would "meet the bill" 80% to 90% of the time.

*Accomplishment*

Here I have been able to find some figures. The inventory of pressure regulators for air and water was reduced from $3,742 to $1,150, a reduction of about two-thirds, saving working capital and storage space and costs. The inventory of valves and cocks was reduced from $12,169 to $4,000, again a two-thirds reduction. Pressure gages were reduced from 27 varieties to 11 (dollar figures unavailable). Bar and sheet stock was reduced from 1,108 varieties to 520. Quotation binders were reduced from 8 types to 2. There were formerly 3 types of transmittal letters going out with instruction manuals, but only 1 was used after standardization. A similar reduction was made in transmittal letters for drawings. Ring binders for our customer catalogs were reduced from 8 types to 2. This is only a partial listing of the reduction we made in item variety, but gives an idea of the possibilities.

### Lower Prices for Purchased Items

As shown above, guiding designers and draftsmen to the purchased items we want them to buy greatly reduces variety. Hand in hand with this goes an increase in quantities of the selected items, and (we hope) lower prices. We felt that in many instances we could eliminate the most expensive types of some materials by standardizing on "good but not magnificent."

#### *Accomplishment*

We did eliminate the "magnificent" purchased items; further, we did achieve many savings as a result of volume purchasing. For example, on pressure regulators the saving was $884 on $8,967 dollars of purchases, or just about 10%. This was for the first year, and it was estimated that savings would be 15% in subsequent years. On valves and cocks we saved $4,998 on purchases of $17,655, or nearly 30%. I found an estimate of $20,500 for the saving in bar and sheet stock, but nothing firm. A saving which was not fully anticipated was the increase in blanket orders. Without standardization our production control department did not dare order very far in advance because of the danger of an item or a size being obsoleted. With standardization we have been able to order a year's supply of many items and obtain an extra discount. We also foresaw and realized savings because purchasing is in a better position to use specifications to seek competitive bids.

### Reduction in Number of Purchase Orders

An attempt was made to find what it costs to make out a requisition, send it to purchasing, then to receiving, make out a purchase order, expedite the order, check the goods on receipt, send the requisition back to purchasing, check the invoice, and make out a check; the total was $6.00 for each order – which is lower than some. You might argue that if we decreased variety and increased ordering quantities so that only one-half as many purchase orders had to be written, the cost per order would go up; and I won't deny it. I will argue vigorously that the cost would not approach twice as much and therefore that we would end up with a substantial saving.

*Accomplishment*

I could find only a few figures. These included: Pressure reducing valves – purchase orders reduced from 75 to 42; valves and cocks – orders reduced from 186 to 95. When fittings were standardized, it was estimated that orders would be reduced from 1,296 to 196!

**Use of Numbers for Identification**

To specify a purchased item fully usually requires a paragraph of written material, and sometimes a whole page or more. With standardization we could write the paragraph or page once, and then need only a number to identify the material. Thus a mere 10 digit number substitutes for a whole page of written specifications. For instance, in addition to dimensions of many selected ball bearings, our standards book gives the ABEC tolerances on I.D., O.D., eccentricity, and width. A further dividend is the recommended shaft diameter and housing bore. This book also has graphs for speed, service, and life factors.

*Accomplishment*

The system works well, and each job of standardization saves a lot of repetitive spec writing by the designers – but I have no dollars-and-cents figures on this.

**Better Control of Purchased Items**

It pays to specify the comparatively small variety of standardized items in complete detail. Without standardization we often take a chance on ordering items without complete specifications and hope for the best.

*Accomplishment*

We have just had an example of what lack of complete and correct specifications can cost. We had been ordering gear motors for one of our products for years by the manufacturer's catalog number only. We recently ordered another lot, and they came in wrong. We found that our number had been wrong all these years. The agent knew what we wanted and always gave the proper instructions to the factory. This agent recently left and we got what we ordered – several expensive gear motors, made special – therefore nonreturnable – that we can't use. Had these gear motors been standardized we would have had a spec sheet to go out with our order, and the manufacturer should have noted the lack of agreement between catalog number and description.

I do not have any figures on the remaining items, but I think you will agree they offer valid savings opportunities.

**Greater Uniformity of Our Manufactured Products**

We often incorporate such items as pressure gages and pressure-reducing valves in our products. It used to be entirely possible for our customers to buy a venturi with pneumatic flow control equipment, a liquid chemical feeder pneumatically paced by the venturi, and one or more dry chemical feeders paced by the same venturi. The Builders flow control might have had a Watts pressure regulator, the Proportioneers pump a Hanlon-Waters regulator and the Omega dry feeder a Kendall regulator. Standardization offered us the easiest way to avoid such a hodgepodge.

**Improvement of Quality Without Having to Change a Lot of Drawings and Bills of Material**

For instance, if the manufacturer of a pressure-reducing valve on which we might standardize should come out with an improved model which was interchangeable, we could switch to the new model by changing only the spec sheet to which the standard number applied. All products carrying that number would automatically get the improved item.

**Uniform Performance Data, Available to All**

For instance, some designers would use one ball bearing catalog and come up with one load rating for a particular type of load; others would use another catalog and come up with a different rating.

I have told you why we put a standardization program into effect. As for where it was applied, you have noticed that my examples included just two types of items – those purchased for incorporation into our products, and those purchased for use in our engineering and sales offices. When I worked for Brown & Sharpe from 1927 to 1932 we had a so-called "screw list," which included hardware items made by B & S and kept in stock. In addition to screws, there were springs, collars, handles, and other items that were incorporated into designs over and over and made substantial savings. When we were setting up our standardization program at B-I-F, I was sure that we would include many of our own hardware items in our book. But nearly every time we estimated the possible saving from standardizing our own hardware items, we were disappointed at the potential – and so we went back to working on purchased items, where we found we could save more. In analyzing the cause of this anomaly, I noted that our sales volume is much less than Brown & Sharpe's and our variety of products is apparently greater. Our use of special homemade hardware of each kind is, therefore, enough lower to make standardization of these items marginal.

### How It Was Done

I hope now that you have seen enough similarity between your problems and ours so that you have decided you should initiate standardization in your plant, and want to know how we did it.

Our first thought was that we would need a very talented man possessing widely diverse attributes, with sound engineering knowledge and judgment. For example, this man would be required to look at a product designed by our Builders Providence group, using Everdur Bronze screws 2" long, at a Proportioneers pump using stainless steel screws 1¾" long, and then at an Omega feeder using Allen head screws 1½" long . . . and realize that 2" long hex-head screws of ordinary bronze would be adequate for all three designs and would not require any design changes. He would then need sufficient salesmanship to induce those responsible for the design of each of these products to change their bills of material. Next he would have to have sufficient business sense to set up the transition so we would not be left with any appreciable dead inventory of the obsoleted screws. This man would also be required to know enough about the products of all our divisions to know where to look for other uses of the Everdur, stainless steel, and Allen head screws. (Otherwise he might eliminate them from the three products, but we would find we had to purchase and stock them for other products. We would then end up with four varieties instead of the previous three.)

Sobering thought: if we could agree that any of our people had the necessary combination of engineering ability, supersalesmanship, business acumen, thoroughness, and product knowledge, we would surely also agree that this person or these people were much too necessary in their jobs for a transfer to standardization!

So we took the group approach. Starting at the bottom, most of the work was done by standardization technical committees appointed for specific products. Each committee had one or more representatives from engineering, design & drafting, shop assembly, shop manufacturing, manufacturing methods, and purchasing. (Sales was normally not represented at this level.) If products being standardized were used by more than one division, representatives of the engineering and design departments of all interested divisions were appointed to pertinent committees. Each committee had a chairman, who might be from any of the areas mentioned. A committee was assigned one item, and when that had been standardized, the committee was disbanded.

Some people served on more than one committee at the same time. If a committee stayed in existence very long, the chairmanship was sometimes rotated. Often one or two people would prepare most of the work for discussion at meetings. One of the most gratifying features of our program was the large number of people who proved to be good chairmen or good spade workers. A mainspring seemed to be available on every committee.

The standardization technical committees were directed by and reported to the manager of standardization. He was a good coordinator, and had a lot of that uncommon virtue, common sense. He was not an engineer or designer, although he had enough drafting experience to be helpful. He had a male assistant, and a secretary.

This manager of standardization appointed the various committees, designated the chairmen, programmed their meetings, and acted as their secretary. He had the right to approve or reject any standards proposed, but once they were approved, it was up to him to effectuate them.

The manager of standardization reported to the director of engineering, but we also had a standardization executive committee which met quarterly and kept the heat on the program. The manager of standardization was given the responsibility of selecting items to be standardized, but these were usually discussed in our quarterly meetings, and only items on which we agreed the opportunities were good were approved for action.

The quarterly meetings were attended by the president of the company, the managers of our division, the engineering vice president, the manager of manufacturing, the manager of standardization and the assistant to the president. This may seem like a lot of brass to put on standardization. To justify it, consider that for all their lives at B-I-F our designers had been free to pick whatever item they wanted for any immediate purpose. We were integrating three companies whose people felt they had equal rights to select and specify materials. Supervisors, of course, sometimes altered decisions, but we had to break long-standing habits. We had to have cooperation throughout the whole company. If any substantial group ignored our standards, we could end up with increased costs instead of lower ones.

**Summary**

Experience shows we developed a sound standardization program. It shows that if your problems are similar to ours, you should expect this from a sound standardization program:

1. A saving in time of engineers, designers, and draftsmen
2. A reduction of inventory
3. Reduced prices paid for purchased items
4. A reduction in the number of purchase orders
5. Designation of purchased items by numbers, each of which can represent from a paragraph to 2 pages of specifications
6. Better control of purchased items
7. Greater uniformity of manufactured products
8. The ability to adopt improved components without changing a single drawing or bill of materials, and to be automatically certain all products will benefit thereby
9. Uniform performance data located in one source available wherever needed (if one copy is missing, another will be close at hand)

# VALUE OF PROPERLY PLANNED STANDARDIZATION

Everett Woerter

Gentlemen:

I have been asked to talk to you about the application of standards in a large diversified engineering and manufacturing concern and the benefits that may be derived from a properly planned standards program.

The necessity for standards, and their value, are recognized throughout all industry today. They are of particular importance to companies engaged in highly technical engineering and manufacturing of the type in which American Machine & Foundry Company is engaged.

An adequate program of standardization coordinated on a company-wide basis offers lasting benefits and tangible financial savings, far in excess of any cost of development or administration.

For many years standardization was the exclusive property of the engineering profession. Today it is recognized that the same principles and techniques can be applied with equally beneficial results in virtually every phase of industry and business: product design, methods engineering, production management, packaging and shipping, purchasing, inventory, and storekeeping. More and more the initiative for setting a standardization program in motion is coming from these departments as realization grows of the corollary benefits of greater availability, simplified maintenance, increased productivity, lower costs, fewer items to carry in stock, lessened hazards of obsolescence, and the like.

Standards, in order to be effective, should reflect what can be accomplished rather than what should be done. They define our present state of knowledge and they live only so long as they continue to define adequately. Standards help to coordinate the efforts of separate groups that may be involved in different phases of the same problem and they promote consistency in the handling of repeated functions while providing answers to repetitive questions.

Standards eliminate the confusion that arises with differences of opinions when consulting various sources of information. When inserted in standards, the evaluation is made but once.

As a result of standardization you may anticipate the following benefits. Standardization enables the buyer and seller to speak the same language and it compels the competitive sellers to do likewise. In addition, by putting vendors on an easily comparable basis, it promotes fairness in competition. It lowers unit costs to the public by making mass production possible. By simplifying the carrying of stocks it makes deliveries quicker and prices lower. It eliminates indecision, a prolific cause of inefficiency and waste both in production and utilization. Standardization helps to eliminate practices which are merely the result of accident or tradition and which tend to impede progress and development.

I would like to explain briefly the AMF organization, so that you will be aware of the extent of our standards effort and the diverse products and companies involved.

AMF consists of 40 plants and 16 engineering laboratories in the United States and Canada. While most companies can enumerate on one hand the items they manufacture, AMF classifies its output according to products for consumers, industry, defense, and atomic energy. It is necessary that our standards be adaptable to the manufacture of atomic research reactors as well as automatic pinspotters for bowling alleys and pretzel bending machines for bakeries.

The majority of work at AMF is of a proprietary nature and the standards effort is governed accordingly. However, some of our locations work very closely with various government agencies on projects concerning engineering, manufacturing, and preparation of standards and performance specifications.

Our Alexandria Division is performing a series of standardization actions for the U. S. Army Signal Equipment Support Agency. Included are simplification studies, specification preparation, and standardization studies.

The standards program at AMF is under the direct supervision of top management. The Standards Department is part of the Engineering Services Laboratory and answers to the vice president of engineering of the corporate organization through the director of research and development. See figure 1.

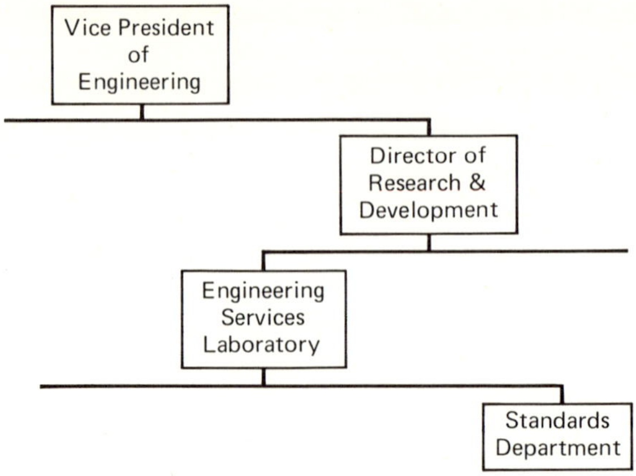


**Figure 1.** Position of Standards Department at American Machine & Foundry Company.

AMF standards are a result of committee action. To insure the formulation of standards that are of equal value to each of the many AMF divisions and subsidiaries regardless of product line, proposed standards are reviewed by a joint standards committee. This committee is comprised of engineering and manufacturing representatives from each of the AMF facilities. The committee's operation is based on the premise, "there is no value in standards merely for the sake of standardizing." Standards have value only if by establishing them you may expect results in economy, when you can simplify operations, or when you may safeguard persons or property. To say this in another way, standards have an actual value only when the money invested in them is returned with interest.

Since the standards program at AMF is a joint effort, the standards committee holds meetings on alternate months, with a plant or subsidiary in a different location acting as host for each meeting. This gives the committee members the opportunity to see the operation of other AMF subsidiaries. As a result, members gain a much better understanding of the overall standards effort at AMF.

Standards committee meetings are conducted as conferences with the manager of the standards department or his designee acting as moderator or conference leader. A request for a standard may be originated by any person, department, or location by submitting the pertinent information to their local standards representative. He in turn evaluates the information submitted to him. If he agrees that the request is valid, he submits it to the joint standards committee for its consideration. The committee then attempts to ascertain the need for a joint standard by asking the following questions of themselves; are the needs presented of a recurring nature – will such a standard be practicable – and, finally – will it yield a profit.

Having determined the need for a standard, the committee then establishes the scope and definition. The sponsorship is assigned to a committee member who is familiar with the subject and is best qualified to prepare a tentative standard. If the committee decides that the problem is of exclusive interest to a particular plant or operation, it is recommended that the local standards committee make an investigation with view of issuing a local standard.

Each location or plant has its own standards committee which is concerned only with problems pertinent to its own operation. Standards prepared for local use are reviewed constantly by the Joint Standards Committee. Thus each committee member is aware of the problems and possible solutions of other facilities or companies within the overall organization.

Many local standards with slight modifications have become joint standards throughout AMF. However, there will always be a need for local standards since, for example, fabrication and process specifications for the manufacture of explosion-proof motors as made by the Leland Electric Company will certainly differ from standards and specifications necessary for the manufacture of the DeWalt Power Saw.

The standards department of the corporate organization, with limited personnel and limited understanding of the individual problems, cannot be expected to have the talent or the time to develop standards on the many specialized subjects of the whole company. Therefore, standards are prepared by the sponsors who, while compiling data, may find it necessary to organize subcommittees or consult specialists or companies having specific interest or experience in the field relating to the subject matter being proposed for a standard. Maximum consideration is given to nationally recognized industrial and technical society standards and specifications as well as standards and specifications developed by the government. It is the standards department's responsibility to keep abreast of all revisions and additions to technical society and industry standards. Periodically all AMF standards are reviewed by the Joint Committee and a summary of revisions or additions to external standards is presented to the committee for consideration. In many cases AMF standards are altered accordingly.

AMF management is of the opinion that the final test of a standard is the degree to which it is accepted and used without external pressures. It should be of such obvious desirability and its benefits should be so completely self-evident that it requires no policing.

We printed recently a tentative standard on springs. While developing this standard a member of the standards department worked cooperatively with the head of the manufacturing research department of one of our plants. While compiling spring data they reviewed many, many hundreds of spring drawings that had been prepared by AMF through the years. Various standards of societies and industry were reviewed, and eight spring companies were consulted before our first rough draft was prepared. The draft was reviewed by the committee and suggested revisions were discussed with the spring companies before being incorporated in our standard. Finally, a member of our standards department visited the Hunter Spring Company and discussed our final document with their engineering department. We standardized all information to be shown on spring drawings and appliques were printed listing only information necessary to properly define a spring. The use of appliques assured that extraneous information would not be entered on drawings. The information that the shop has to work from is always similar, and presented to them in the same orderly fashion.

The AMF Pinspotter Division is an excellent example of standards put to use. The Pinspotter Division as such is an engineering, purchasing, and assembly operation. It does virtually no manufacturing. Parts and components are manufactured in AMF plants in Buffalo, in Brooklyn, and by many subcontractors throughout Ohio. During peak production they are turning out approximately 1,200 Pinspotters per month and during the slack period approximately 400 machines a month. This represents reasonably high production of a complex machine.

The use of standards in the design stage allows for uniform interpretation of drawings by all concerned. Standards also facilitate manufacturing and inspection and insure proper fit and interchangeability of parts. Thus, good product design, modern manufacturing and assembly techniques, plus proper application of standards has enabled AMF to market their product to the point where they recently signed a lease for their 30,000th machine.

It is rather difficult to determine savings directly attributable to a standards program. In any program the benefits received from the use of technical standards are both tangible and intangible. Realization of the objectives of adequate standardization is a part of the net return. Some standards are essential for modern manufacturing operation while others promote efficiency in a more direct way. In any event specific valuation is difficult and presumably must be based in each instance on an estimated comparison of costs with and without the standard.

Approximately five years ago AMF formed the corporated standards department to coordinate standards on a company-wide basis and eliminate duplication of effort within the AMF family. Shortly after our organization, we took inventory of our accomplishments and used the figures compiled for the year of 1952 as a yardstick. Since no records of the dollar value realized through the use of technical standards were known to exist in AMF, evaluation was based on *opinion* only. However, in order to present the importance and magnitude of this subject, based on the comparison methods suggested above, the following figures are offered as being *conservative opinions*.

At the time of our investigation, the standard we had prepared on drafting room practice revealed the following information. The replacement of delineation with description for simple parts such as pins, tie rods, studs, bearing retainers, shims, compression springs, spacers, washers, etc., would have saved a minimum of $2,400 per year in Drafting Room "A".

It was estimated that drafting room practice including simplified drafting would have saved up to 40% of total design time, which in the case of Drafting Room "B" would have represented 33,000 hours or approximately $165,000 a year.

It was pointed out that the processing of drawing changes in Engineering Department "B" cost an average of $30 per change. The total expenditure for this purpose was approximately $60,000 per year. It was estimated that the standard on this subject would have cut this time in half with savings of $30,000 per year.

Our investigation showed that machining tolerances were an excellent source of cost control through education by engineering standards. A summary shown in table 1 of relative costs of various commonly used tolerances indicated the possibilities of savings.

The improper specification of tolerances can increase the cost of the operation up to 1,700 percent. Practically all machining operations are dependent upon tolerances. Being conservative and permitting but one error in 200, the increased cost of manufacturing at Plant "C" amounted to approximately $24,000 per year.

**Table 1. Cost of Various Tolerances Compared with ± 0.030″ Tolerances**

| *Tolerances (Inches Plus or Minus)* | *Relative Cost* |
|---|---|
| 0.001 | 17 X 1 |
| 0.002 | 11 X 1 |
| 0.005 | 5 X 1 |
| 0.010 | 3 X 1 |
| 0.015 | 2 X 1 |
| 0.030 | 1 X 1 |

The standard on cylindrical fits would have saved Plant "D" up to $50,000 per year.

The standard on tolerances including parallelism and squareness would have saved this same plant approximately $6,000 per year on Pinspotter parts alone.

When preparing a standard on surface finishes, it was found that engineering standards that have a direct bearing on manufacturing costs are extremely important. At the time of this investigation one of our manufacturing facilities (Plant "C") was scheduling better than 12 million dollars worth of machining per year. Call-outs for machine surface finishes, tolerances, and material specifications were three items that affected manufacturing costs to a considerable degree.

Machine surface finishes were specified in root-mean-square value of roughness in microinches in order to provide for greater control over surface finish requirements. The differences in cost of a few of the more commonly used finishes were listed as shown in table 2.

**Table 2. Cost of Surface Finish at 500 Microinches RMS Roughness Value Compared with Other Roughness Values**

| *Microinches (RMS)* | *Relative Cost* |
|---|---|
| 16 | 3.7 X 1 |
| 63 | 2.3 X 1 |
| 125 | 1.8 X 1 |
| 500 | 1.0 X 1 |

Calling out a finish that was finer than absolutely required increased the cost of the machining operation anywhere from 28% to 370%. Obviously, proper guidance through standards was required under the circumstances.

Machined surfaces consumed roughly 2/3 of the total fabrication time required for producing an average part. This left approximately 8 million dollars worth of machining time dependent upon surface finish specifications. This is when considering the 12 million dollars total booking for the year. The slightest mistake in specification of surface finish would increase the cost of operation 28%. Were this error to occur once in every 100 surface specifications, the yearly increase in manufacturing costs would have been $22,400.

It was estimated that one of our facilities (Plant "D") was spending in excess of $10,000 per year producing better finishes than those functionally required.

Material costs represented a considerable sum of money and were well worthy of control by the guidance of engineering standards. The specification of a material that was superior to its functional requirements would have increased the cost of material anywhere from 10% to 600%. Assuming that a material was called out once in 25 parts which increased the cost 20%, the yearly increase in material costs at Plant "C" would have amounted to $14,400.

The use of standard parts offered a great deal of possibility for cost reduction. Using sprockets as an example, it was found that one of the important considerations in standardizing was the advantage of buying or manufacturing in large quantities. By standardizing, the number of types was reduced with an attendant increase in number of each type. Table 3 is a short tabulation of costs for varying quantities of two sprockets.

Purchasing or manufacturing in lots of 100 rather than 5 represented a reduction in cost of 57%. The difference between a lot of 25 to 50,

**Table 3. Variation, by Quantity Purchased, in Unit Price of 2 Parts**

| *No. of Pieces* | *Part A* | *Part B* |
|---|---|---|
| 5 | $9.65 | $6.75 |
| 25 | 5.09 | 3.29 |
| 50 | 4.60 | 2.93 |
| 100 | 4.35 | 2.75 |

though a slight change, represented a 15% reduction in cost. The yearly reduction in cost in Plant "C", considering that 60% of our sprocket production would fall into standard parts, would have been approximately $4,400 per year. In addition, savings in capital invested as well as storage space would have been realized.

Standardization of commercial parts would have allowed us to –

1. Mass-produce standard manufactured parts for stock in large quantities
2. Permit company-wide utilization of costly patterns and dies
3. Reduce engineering hours required for delineation of standard manufactured parts
4. Purchase commercial parts under a consolidated purchasing technique which would permit company-wide volume purchasing

During 1952 Design Section "B" was able to reduce its drafting time from 5.2 to 4.3 hours per square foot. It was estimated that at least half of this saving or approximately 0.4 of an hour per square foot was due to standards. This 0.4 of an hour during 1952 represented approximately 6,000 hours or $33,000 saving.

Due to the standard on decimal system of dimensioning, people in our shop no longer have to convert fractions to decimals. It was estimated that this factor alone might save $5,000 per year of scrap.

Our experience at AMF has shown that acceptance of standards is slow in pace and is often hampered by misinterpretations, personalities, and just plain resistance to change. Some of these obstacles were overcome, and others are disappearing. But the overall results have been extremely successful and the program is gaining enthusiastic endorsement and cooperation from management of all divisions and departments of AMF. Extensive reeducation programs are in effect to sell new standards where none have existed before. And, as acceptance increases, wider cooperation between divisions has developed. This has resulted ultimately in vast savings by distributing work loads, both engineering and manufacturing, over several independent divisions.

We at AMF feel we have proved that flexible, adaptable standards, voluntarily adopted by divisions and subsidiaries, serve a large and diversified organization well – and profitably.

# THE ROLE OF MANAGEMENT IN COMPANY STANDARDIZATION

John T. Milek

Industrial or company standards, specifications, and standardization, when viewed in a broad sense, cover a wide field of endeavors and activities in any company organization. We can lump them all into one package when making a survey of a typical company operation. To simplify our survey tour, I have designed a "Railroad Track Circuit" to make the rounds into each department of the company.

The various departments are shown as stations along the route; the customer is shown on a railroad spur track; and management is in the control tower, throwing the switches to route the trains. Management also gives the red or green light to standards activities in the various departments. The railway cars contain the purchase orders, specification documents, standards documents, blueprints, and drawings.

We can also visualize the circuit as a chain. You are acquainted with the well-known cliche, "A chain is no stronger than its weakest link" – a weak standards operation in any department can give a company trouble.

In 1936, John Gaillard likened this management-standards department relationship to a symphony orchestra conductor and the orchestra itself. All the instruments have been tuned to a standard pitch and they play a common musical standard language – the score.

## Accounting Department

Let us start alphabetically with the accounting department and see what standards and standardization we can expect here.

Standard costing gives us a benchmark against which we can measure our actual costs and thereby determine the quality of our performance.

If accounting records contained nonuniform item identifications, serious problems would prevail. Correct nomenclature clarifies records and bookkeeping to a large degree. One company, for example, states that before standardization, common, ordinary washers were masquerading under a total of 57 different names: ring, bushing, collet, grommet, spacer, etc.

Data processing machine techniques require some standardization of forms and operations, symbols, language, and nomenclature to handle information or data. The use of standard IBM cards brings enormous savings in processing time as they fit standard IBM machines.

**Administration Department**

Some firms have developed administrative standards for training. These involve management duties, responsibilities, and authority, and are spelled out in CSP's or company standards practices. The American Management Association recently conducted and published a management study titled "Setting Standards for Executive Performance." The study points out the difficulties in establishing standards in this area of company management.

**Advertising Department**

This department can make extensive use of standards and specifications in its product line advertisement.

In its advertising literature, for example, the fact that a company's product meets the requirements of national, industry, or military standards can be an effective selling point.

Quoting a short paragraph from American Standards Association literature: "Advertisers are discovering a simple truth – that an American Standard is good value. Advertisements which feature compliance to American Standards and other nationally recognized standards tell prospective purchasers that products meet requirements agreed upon by users themselves as well as by manufacturers, distributors, and technical experts."

**Contracts Administration Department**

We find a contracts administration department in a company when it is engaged in government or military contract work. The contracts generally set forth the requirements for the equipment (e.g., performance, operation, reliability), drawing requirements in accordance with certain MIL specifications or MIL standards.

A contract further indicates which general equipment specifications are applicable in a particular case. Examples are MIL-E-5400 covering airborne equipment, MIL-E-4158 and MIL-E-16400 applicable to ground-based equipment. These general specifications list federal, military, Air Force, Army, Navy, and miscellaneous other specifications and publications which must be followed in the selection of mechanical and electrical parts, materials, finishes, and in the preparation of drawings and service manuals. The individual equipment specifications include more specific information with regard to service life and shock and vibration requirements.

In some companies, it is the function and responsibility of the contracts administration department to analyze these specifications, to determine whether in the manufacture and design of the end equipment the specifications are conformed to, and to see that they are.

### Design and Engineering Department

The design and engineering department should be among the most aggressive supporters of a strong standards program. As a department they favor standards, but as individuals they may, at times, resist standards as limiting their scope or restricting their ability to come up with the precise design that best meets the technical requirements. I would like to stress that the role of standards should always be considered as an aid rather than as a restriction to design engineers. Designers should be urged to consult and apply standards in their designs in the same manner that they would use established mathematical and physical formulas.

Use of standard parts in these same designs assures the design engineer lower purchase price and increased value. Service and repair parts are easier to obtain, and the use of standardized processes in the manufacturing of the part further assures him of better original quality. Company standards should become part of the design engineer's daily working tools, and should be merged with his other daily implements of design.

Drafting standards can save thousands of man-hours in the drafting section by whittling down costly drawing changes. Standardization of drawings and drafting-room practices saves many thousands of man-hours annually. A new technique in this area is the microfilm aperture card developed by the Department of Defense for its contractural activities and which can also be utilized by companies for their own extensive blueprint or drawing files.

Simplified drawing has been proved to save dollars and man-hour effort, although it is still the subject of debate and argument in drafting circles. The design draftsman must know company standards, standard materials, standard hardware, preferred sizes, shop-run tolerances, and manufacturing processes. Lack of knowledge in any of these things can be costly. The use of preferred numbers, American Standard Z17.1-1958, can save a company many thousands of dollars over the years.

Training of newly hired engineers in the use of company standards, manuals, specifications, and drafting practices to obtain a comprehensive picture of a company standards and standardization program is emphasized by many large company organizations, even to the extent of making short-run motion picture films to introduce these facets of the company operation. Two companies in Southern California who have done this with great success are Convair-Pomona and Librascope, in Glendale.

Several case histories involving engineering design and standardization are quoted here, from the American Standards Association brochure titled *Dollar Savings Through Standards*.

Case No. 1 involves the Denison Engineering Company, Columbus, Ohio, which manufactures hydraulic equipment presses, controls, and power units.

> When designing new products and redesigning products, Engineering is incorporating JIC (Joint Industry Council) Hydraulic Standards in addition to Machine Tool Standards and other industrial standards so that our customers will have products with the latest features at no additional cost. In the past, it has been necessary to charge extra for these features on products that had already been in production. This is still true of some of our older products.
>
> We find that by designing and incorporating the established standards there is a benefit to us and our customers, as long as these standards do not restrain design ingenuity to the point that design progress is affected.
>
> We believe that a further standardization in the hydraulic equipment group would enable the purchaser of machinery and hydraulic equipment to derive ultimate savings in cost and reduction in service parts.

Case No. 2 involves the XYZ Company which is primarily an engineering organization engaged in the design and construction of petroleum and chemical plants.

> Each such plant is specially designed to perform a particular specified service. In other words, we produce a custom-built product which cannot be standardized because of the diversity of purposes and conditions pertaining.
>
> During the engineering, designing, purchasing, and construction stages of each product, company, industry, and association standards are widely used. In so far as possible, procedures and data for engineering are standardized. Integrated parts such as pumps, instruments, motors, etc, are standardized to reduce varieties as to size and type and to avail ourselves of parts readily available and within the standards, specifications, and codes of customers, associations, and states. In construction, methods of erection are standardized to a degree, and our construction equipment is standardized somewhat to reduce varieties and types and parts therefor, which have to be procured or stocked.

Case No. 3: The Monsanto Chemical Company stated in 1955 that they saved, through a standardization program on pumps, about $64,000 as follows: $15,000 in capital, $5,000 in inventory, $12,000 in maintenance, and remainder in design time which amounted to $32,000.

A few problems crop up when a "special" product design is ordered:

1. New cost estimates must be made.
2. Special handling of the quotation by the estimating department involves extra man-hours of effort.
3. Extra time for the engineering department upon receipt of the order.
4. Extra set-ups on machines in the plant – special jigs and fixtures.
5. Short production runs.
6. Multiple stock of finished parts.
7. Longer deliveries which result in customer dissatisfaction.
8. Disruption of normal scheduling and production.

The following recommendations are offered to engineering designers:

1. Consider standards – design or modify so that available standard parts or materials may be used. There are a wide variety of standards and specifications to work from: commercial, industrial, national, MIL, federal, ASTM, SAE, American Standards, international, etc.
2. Design around standard fasteners, washers, eyelets, spacers, bushings, terminals, tubes, structural members, etc.
3. Use industry-wide standard materials furnished by specialty vendors.
4. Specify standard sizes for raw materials to avoid extras in cost.

### Maintenance and Construction Departments

In some companies, construction standards for plants are a requirement, especially in a petrochemical or oil refinery type operation. Here is required the following: Engineering specs for plant layout, piping arrangements, pumps, fire protection water system, portable fire extinguishers, design of electrical systems, and design of gas pipeline systems.

One company (Pan American Petroleum Corporation) reported that, prior to standardization, over 60 drawings of Christmas trees were used in the company. In reducing the number of trees, three basic types were established according to working pressure, and each type was designed so that alternates could be used to meet particular operating conditions. In addition to standards for Christmas trees, they established a standard arrangement of equipment for blowout preventers, work trucks, wellhead hookup for injection wells, and installation of instruments.

Many other examples can be cited where standards and standardization can be effective from a maintenance or construction operational point of view:

1. Standard maintenance forms and methods of reporting malfunction of equipment
2. Building widths, uniform column spacing, and entrance for machinery
3. Layouts for power service to meet growth and rearrangement needs
4. Standard color schemes for office and factory painting
5. Standard office and factory partitions
6. Standard industrial lighting fixtures
7. Standard furniture and fixtures

### Manufacturing and Production Department

The manufacturing department is usually a strong supporter of dimensional and materials standards, as these result in larger manufacturing runs and closely defined limits of materials properties. However, it may, at times, find standards a constraint on its freedom and on its flexibility to meet schedules. Also, at times, it may vacillate between tight or loose standards, depending on its immediate interest.

The various standards and standardization areas and activities possible in the production engineering functions are the following: Process standards, methods standards, tool standards, fixtures, gages, materials handling standards, plant layout standards, time study and work measurement, machine tool interchangeability.

Shop practice standards cut drafting time, errors, and costs, and assist the designer, draftsman, the shop personnel, quality control inspector, and purchasing. These standards generally include or cover: Flatness, straightness, parallelism, squareness, angularity, roundness, taper, concentricity, alignment, edges and corners, and surface quality.

Other supplementary requirements are surfaces produced by machining or casting; surfaces produced by press brake operations on sheet metal; threads and tapped holes; cylindrical surfaces (internal and external); drilled holes; slots and keyways; radii, chamfers, burrs, and undercuts.

Great savings in a manufacturing plant are effected through the use of standard motors, using the American Standard (National Electrical Manufacturers Association) frame sizes which result in uniform dimensions regardless of who makes the motor.

Another plant example is lubrication fittings. A number of companies reported having special fittings purchased at increased prices. In one company, 90 different fittings accounts were carried; upon standardizing, 40 accounts were closed.

Another company said that standardizing on antifriction bearings saved many dollars by not stocking duplicate bearings under different numbers. Standardization resulted in savings of space, record keeping, and inventory value.

## Purchasing Department

Invariably, the purchasing department is a strong supporter of standards. They have direct evidence of the many advantages that result when they can purchase standard items, and the penalties when they require a nonstandard item. The National Association of Purchasing Agents has long recognized the importance of standards and has done much to publicize and promote such work.

In one company, the purchasing department reduced 30 different paints to 15, 120 different cutting fluids to 10, 50 different tool steels to 6, and 12 different aluminum casting alloys to 3.

H. W. Christensen, Director of Purchases, Columbia-Geneva Steel Division of U. S. Steel Corporation, in the December 1957 issue of *The Magazine of Standards*, described the following standardization benefits as realized in his purchasing department:

> Among the benefits was the elimination of 22 percent of items in several commodity classes because of duplications and obsolescence. Use of standard item descriptions, nomenclature, and industry standard numbers made it possible to reduce purchase costs of many items from 30 to 40 percent. Adoption of industry standard numbers gives flexibility in making purchases.

He said further, "We are convinced by demonstrated performance that these programs result in cost savings, efficiency, and progress." However, he warned that standards should be flexible enough to permit progress or modifications since we are in the midst of a revolution in materials due to industrial and technological developments.

It is not often recognized or appreciated that purchasing men can buy plant maintenance equipment, supplies (such as paints, solvents, oils, lubricants, and greases) per MIL, or federal specifications, or nuts, bolts, rivets, washers, etc, per National Aeronautical Standards and save company money.

Harlan E. Cross, purchasing agent of the U. S. Pipe & Foundry Company, Birmingham, Alabama, in the December 1957 issue of *The Magazine of Standards* cited the following areas where a lack of standardization contributes to inefficient buying and loss of profit:

1. Excessive stocks of materials and replacement parts being purchased.
2. Slow movement and stagnation of certain stock items.
3. Time lost in often fruitless searches for items which alone will serve a need.
4. Private stocks accumulated by departments.
5. Too great a number of varieties and sizes of materials.
6. Excessive number of trade-named items purchased rather than purchasing on a specification basis (more will be said about this later).
7. Lost motion and confusion in the requisitioning and purchase of materials.
8. Too great a number of specially designed items being purchased.
9. Lack of uniformity in the terminology used to describe the items.

## Quality Control Department

Quality control is also a strong supporter of a standards program. It is the enforcement agency of company standards. It is the duty of quality control to act in an enforcement capacity to ensure conformance to company standards.

Quality characteristics are usually measurable in some form or other. Lord Kelvin once said, "When you can measure what you are speaking about and express it in numbers, you know something about it; but when you cannot express it in numbers, your knowledge is of a meager and unsatisfactory kind."

As a general rule, engineering can effectively translate standards into numerical figures. It is a part of standardization to determine the method of measurement in order to avoid the manufacture of nonuniform products. Uniformity of practice and of product can be destroyed by sloppy and nonstandard gaging. In the field of measurement, the standards engineer plays an important role. Planning a job must include planning the gaging. Gaging responsibility ultimately lies with quality control.

It should be recognized that the intelligent use of standards and specifications is essential to good quality. Close liaison between standards engineering and quality control is a must, for mass production demands interchangeability and only intelligently controlled standards will bring about this goal.

## Reliability Department

Reliability is a new concept that has crept into the industrial picture, stemming mainly from the military. The exact relationship between reliability and standardization is difficult to describe in a few words. Moreover, the military have, in the past two years, prepared a number of MIL reliability specifications:

1. MIL-R-26667 "General Specifications for Reliability and Longevity Requirements for Electronic Equipment"
2. MIL-R-26674 "Reliability Requirements for Weapons Systems"
3. MIL-R-19610 "General Specification for Reliability of Production Electronic Equipment"
4. MIL-R-25717 "Reliability Assurance Program for Electronic Equipment"

Navy MIL-R-19610 outlines the minimum requirements which the contractor must perform on production equipment to assure the production of reliable equipment. Levels of reliability are defined in terms of operating hours without failures. Minimum test requirements for electronic parts are given, and a quality assurance provision for completed equipments is specified. The other specifications have about the same requirements.

Reliability, like safety, is a desirable quality which has often been achieved in the past without serious difficulty. It is something everyone wants but it is hard to define, difficult to specify, and, because of the complexity of modern equipment, is becoming more difficult to achieve.

Because of the widespread realization that electronic equipment reliability is vital to national defense, the military departments and industry have given it a lot of attention. As a result, electronic devices that are significantly more reliable are beginning to reach the field.

Reliability engineering is responsible for the engineering specifications needed to produce inherently reliable equipment. It is also responsible for qualifying sample parts so that they obey the exponential failure law and have an acceptable failure rate for a suitable time when used in the planned application.

Since all reliability, gaging, testing, and screening of parts aim at providing a product which will exhibit a stabilized constant average failure rate for a specified time under specified use conditions, the most direct way to specify part reliability is to state the required stabilized failure rate, the time of operation over which this failure is expected, and the test conditions which will approximate the end-use environment.

Type approval for reliability is performed by the engineering department, and usually consists of life tests being run under end-use conditions on a small quantity of sample parts.

The significance of reliability specification in contracts involves increase in: (a) manpower, (b) equipment, (c) number of parts tested to failure, (d) man-hours, (e) added costs to finished products.

In some companies, reliability is a part of the quality control department, while in others, it is a separate function of the engineering department.

**Research Department**

The research department may consider the use of standards as an unnecessary straightjacket to their work. Research can often be assigned a project that develops the limits of standards for a product line or component.

The market research department on many occasions determines the need for a product or material and establishes some tentative general requirements. Research and development studies the problems of material, configuration, and basic design. Pilot lots are made and tested in the laboratory. Engineering is then called in to incorporate the necessary consideration of standard requirements that may be needed. Market research work is simplified when there are standard industry designations for products.

Some companies have primary standards laboratories for calibration and measurement purposes. Close tie-ins are made with the National Bureau of Standards.

The American Society for Testing and Materials states that all of its ASTM standards are predicated upon an extensive foundation of laboratory research work performed in the research laboratories of its member companies.

### Safety and Health Department

In the formation of standards for safety and health, consideration must be given to the safety of employees and public as well as to continuity of service. Safety departments have available the numerous standards established by the U. S. Department of Labor, National Board of Fire Underwriters, Underwriters' Laboratories, and the American Standards Association.

Many companies must design and manufacture safety and health standards into their products. Examples are children's toys, which must be safe so that children are not poisoned by lead in the painted surfaces.

Safety standards are put into good lighting in the plant and into heating, ventilating, rest rooms, lunchrooms, etc. Other safety standards provide for protective clothing, eye protection, foot protection, and electric shock protection. Safe practices are established with regard to care and use of abrasive wheels, power presses, cranes, derricks, jacks, ladders, welding, and color code for marking hazards.

### Shipping Department

For the shipping department, standard parts mean quicker delivery to the customer and a minimum amount of paper work processing, especially if parts identification coding is invoked. More prompt and efficient service is available to a company's customers, especially on reorders and repairs.

### Standards Department

It is the task of the standards department to cooperate with management in defining the standards activity that is best for the particular business.

What are the various functions of a standards department?

1. Define or develop a policy of standardization for the basic operation of the company and submit this to management for adoption.
2. Centralize and keep up to date a collection or library of outside publications on standardization, especially national standards, industry, commercial, military, federal, and all technical information which qualifies or supplements them.
3. Circulate the national standards to the departments which may be able to adopt them if the need arises. These standards may also serve as a reference to all engineering functions.
4. Establish internal standards publications for various sections of the company.
5. Keep track of the application of national standards and internal standards in various sections of the company.
6. Participate in or be represented on national standardization committees, as well as national delegations sent under authority of the national standards activity to the international standardization meetings.
7. Address standards publications inquiries to the national standards body, and be sure that the company position is defined and transmitted to the cognizant trade organization when an industry position is to be resolved and submitted to the appropriate national standardization committee.
8. Circulate, within the company, data on reduced costs and other benefits obtained through standardization.
9. Act as a clearing house for the company's engineering standards efforts in addition to preparation, coordination, and maintaining of company standards.
10. A standards department can assist in the task of identifying items acquired and used by the firm in its operations. Materials and parts must be numbered, named, or identified by combinations of these and other means. The finished product, its components, and the supplies and equipment utilized in its manufacture must retain their identity at all times. Most standards groups have a copy of the *Federal Item Identification Guide* or

the *Federal Catalog System of the Department of Defense*. The book is distributed by the Armed Forces Supply Center, Washington, 25, D.C., and represents a monumental task of item identification which many companies can use extensively in their operations.

The standards department should bring to the attention of management new standards developments, programs, or projects of other companies which have rendered savings, such as the following examples.

The International Business Machines Corporation has developed a new type of standards program in which a new data classification system makes it possible to arrange material in the standards book for the benefit of the user and also to meet the requirements of mechanized data processing. Electronic data processing equipment is applied to the problems of preparing standards through cataloging, analysis of trends, and prediction of future requirements.

The DuPont Company claims that their company's standards program is resulting in a return of about $4 for every dollar spent on standards activities.

W. F. Rockwell, Jr, president of the Rockwell Manufacturing Company, cited some standardization savings made by his standards department in the June-July 1959 issue of *Standards Engineering*:

> Two of the plants, working together, analyzed 123 types and sizes of hex nuts which had been standard inventory throughout the company. They were able to reduce the 123 to 54, and of these only 27 retained their original specifications. In these two plants, on this one simple item, the resulting saving in cost of nuts used and in assembly time is at the rate of $5,000 a year.
>
> The standards department usually deals with methods and materials for the repetitive and the routine calculations and uses that the engineer is concerned with, leaving the engineer free to apply originality and thought to the nonrepetitive, complex aspects of his work. To cite an example:
>
> Several years ago, Federal Specification QQ-B-611 on brass was superseded in part by QQ-B-613 and in part, by QQ-B-626. The new specifications use different designations for composition and cover a different group of available forms and conditions. Here is where, subsequently, purchasing agents, vendors, and inspectors turn green and where rigor mortis sets in if specifications are not reviewed regularly and records kept up to date.
>
> The designer on the board certainly cannot keep abreast of such changes and must therefore rely on an alert, competent standards group to keep him abreast of the latest specification revisions, designations, and perhaps even material characteristics, and screw thread standards.

A specific case history of what can be involved in the preceding was detailed several years ago by the Fairchild Camera Company which, during a period of two months in 1955, reviewed 1,192 drawings connected with their production reorders. Sixty percent required changes in the material, finish, or process specification number, exclusive of amendment letter changes. This required more than just drafting time; in some cases, the specification review involved small-sized safaris through jungles of overlapping specifications.

An active, aggressive standards department should provide a design guide to the various other departments and contain the following:

1. Individual equipment specifications
2. General equipment specifications
3. Mechanical design standards and modular design
4. Lists of preferred parts and preferred numbers
5. Lists of hardware (screws, nuts, washers, knobs, etc)
6. Lists of standard materials for selection of metals, ceramics, plastics, adhesives, etc
7. Lists of standard finishes and processes
8. List of wiring standards
9. List of mechanical assembly standards

**Warehouse, Stockroom, and Receiving Department**

Occasionally, standards problems can occur in a warehouse, as demonstrated by the following example:

A given grade of copper-silicon rod stock may be procured and stored under any one of the following standard designations: MIL-B-15939; ASTM B-150, Alloy No. 1; SAE 701-B; AMS 4632B; Federal Specification QQ-B-666, Grade B; Army-Navy-Aeronautics Specification AN-B-11; Navy Specification 46B17, Grade B; and many proprietary or trade designations: aluminum bronze, Ambralloy 928, or Mueller 224-E.

Should the metal be purchased at different times under several of these standards by different engineers in different departments, each one of the purchases might be carried in the stockroom and on the records as a separate item. Standardization takes on meaning when the metal is designated uniformly throughout the company and handled in a standard manner.

Coordination of stockroom and warehouse with the standards department can effect considerable savings.

This completes the tour of our outer railroad circuit.

### Customer

In all industrial standards work, we are interested in the product – the product which goes to the customer. With standards, the customer is buying the best article for his money. Standard products are supplied much faster to the customer than "specials," which require disruption of production, diversion of labor effort, special tools, special techniques or processes, special materials, etc.

When the customer is the federal government or the military services, we have to deal with standards and specifications on a high level of operation. In many cases, the government sets standards for the sole purpose of protecting the public – the individual consumer, John Doe. Without a doubt, the federal government is one of the greatest compilers of standards.

It is the customer with his purchase order and specifications that sets our train into motion, bringing the blueprints or drawings into each segment of the company operation.

Let us finally look at the main control tower of our railroad system – management.

### Management

It is essential that management have a clear-cut viewpoint regarding the objectives assigned to the standards department which it has created and the objectives which it hopes to obtain.

The definition of standardization policy is essentially a function of management, and the standards department is, in principle, the executor. We must recognize that some difficulty can exist in defining what the standardization policy should be in some companies. Each company has its own policy, and it must be the primary concern of the person or persons responsible for standardization to analyze precisely and practically the goals which he or they hope the service will attain.

Problems facing management regarding a standard operation are:

1. How is the standards effort to be organized?
2. How big an effort should it be?
3. How are the results of the effort to be measured?
4. Should the standards effort be a staff or line function?

5. To whom should the standards manager report?
6. What type of standards organization shall be adopted?
7. What areas shall be covered by the standards program?
8. What shall be the operating procedure?
9. What shall be the level of a standards operation?

It must be recognized that a standards program is only as strong as the support it receives from company management. Management support of the standards program perhaps can be best expressed in the form of a managerial standard, approved and signed by a top company official, such as its president or general manager, that defines the operating procedure for the standards program.

Management must oversee and follow up if standards are to be recognized, put into force, kept in operation in each division, department, section, or group of its organization, and also kept dynamic to meet future needs and changes in the industry, consumer requirements, and customer satisfaction.

Standards are a management tool; standardization is a management function.

Management should be interested in standardization because it can make and save money for every company in business.

# LET'S TALK SENSE ABOUT COMPANY STANDARDS

Arnold M. Rosenwald

The consumer is passing out prizes today at a rate never before experienced in the history of commerce. Not only are the American public and corporate consumer buying more, but the market today is truly global. What is this prize the consumer awards? He selects the product of your company in preference to those of all of your competitors.

Why does he? We have to assume that everything offered for sale will perform the function for which it was intended, otherwise the producing company has goofed with large numbers of dollars. The motivation research and marketing people can come up with a good many reasons why companies and people buy the way they do. I don't intend to discuss all of these reasons because they are altogether another subject. However, I do want to emphasize one important factor contributing to the consumer's choice and that is the price tag you put on your product.

Many factors enter into the pricing of an item and most can be lumped under the general headings of (1) material and (2) labor performed on the material. Perhaps you answer: "To perform a given function we have to provide an item that will do thus and so. My competitor has to do it, too, so our pricing must be in the same range." This last thought is precisely where two companies may begin to draw apart, when one operates efficiently and the other not quite so well.

What contributes to efficiency? To name a few principal contributors, your material must be the most efficient of all of the available materials; your fabrication procedures should work that material efficiently; and your people have to devote their time to production of the product, and also to planning for design and production of new products, with as little lost motion as possible.

This thinking applies equally well to the production of toasters, ten-ton cranes, or barrels of plastic molding compound. All these items have this in common: they are materials formed into a useful shape and offered for sale to satisfy a need. How we "oil" up the company machinery by the modern management technique of company standardization is the target of this article.

### Definition and Purpose

There have been many highly technical and distressingly long definitions offered for a standard, and particularly a company standard. I prefer to think of the company standard as being simply an agreement among those in the company having substantial interest in the topic as to how they will handle a situation that comes up frequently enough to be bothersome. Thus you have laid out a preplanned course of action to prevent troubles or waste. You have freed creative talents for the special aspects of every job that comes up which in fact make it different from the last job. In short, the company standard must be useful and it must promote communication in recurring situations.

As a company grows and gains experience, this experience must be recorded in order that it may have lasting value. Otherwise, accumulated experience is available to the company only as long as the individuals who gained it are still working in the company. The labor turnover rate of the sixties makes the retention of company know-how a major consideration. By your development of the company standard, you insure that yesterday's mistakes will not occur tomorrow if people apply this standard. And in the positive sense, what you did correctly yesterday will again be done correctly tomorrow and, hopefully, a little better. One major function performed by the development of a company standard is to provide the basis for further improvement in techniques. No company standard can be considered permanent once it is finished. We must update our standards as we learn more about what we are doing.

### Reasons for a Program

There are some rather simple and yet startling reasons for a company standards program. Consider this: managements exercise control on whom they allow to buy parts or services or materials, or to sign purchase orders. They keep checks locked up in the accounting department in order to prevent forgeries and other misuse. But they let countless people with varying levels of know-how specify manufacturing requirements, exotic materials, and special parts on the product drawings or specifications, which commit company dollars just as writing a check finally releases those dollars. In effect, what happens is that even the most skillful purchasing agent is tied down by arbitrary limits beyond which he cannot negotiate further price reductions. Thus under the guise of "complete design freedom" more can be spent in hidden costs of production than can ever be retrieved via a competitive selling price.

Consider another aspect of the lack of company standards. Many companies have national and world-wide distribution facilities. They back up all of this with a far-flung service organization. As a matter of fact, the volume demands of much of today's business require just such a large scale operation as this. If a few designers in the engineering department have complete freedom to select parts that go into the product line, and three or four select as many different parts to perform the same function, then the supply pipeline is necessarily larger and embraces more items. Large-scale stocking of several duplicated items is required, rather than smaller stocking of only selected parts. With shipping, warehousing, and distribution costs what they are today, this can all add up to a back-breaking expense in the supply picture. Consider also that if assemblies functioning wholly differently are used in a company's line of equipment to perform the same overall task, considerable additional training of servicemen is needed to provide a customer with the service that he demands at a reasonable price.

Each company has to make its own decision as to how much importance it will assign to different topics. However, consider that the designer may improve the serviceability of a product by thinking about it, but can seriously impede and increase repair costs by giving serviceability third or fourth place in the order of importance. I need only remind you of where sparkplugs were located on the V8 engine of a few years ago as a result of emphasizing a low hood line to the exclusion of other considerations. In this case, the manufacturers no doubt felt that if the car was not highly styled, it would not sell, and if it did not sell, there was no service problem. There are signs today that designers are learning how to design for serviceability as well as for high styling.

## Standards Improve Performance

Once you have found out how to do a job well, it is wasteful, if not downright stupid, to degrade performance. This is exactly what happens, though, when you gain experience in designing an item and then put a new designer onto the task for next year's model without preparation. When he is not familiar with all that went before, he practically goes back to scratch in the design evolution.

To illustrate, I owned a 1951 model of a particular make of car. The defroster was very effective in that, as soon as the engine approached operating temperature, I could depend on a quick defrosting and defogging on my side of the windshield. When I traded this car for a 1955 model from the same manufacturer, I found that the defroster was just as efficient but that the passenger's side of the windshield would clear quickly, and the driver's side would remain clouded for some time. The problem was that in the intervening years, the defroster fan was moved to the right side of the windshield from the driver's side. Thus, poor performance resulted from neglect of proven performance. In discussing this problem with a friend who had previously owned a 1950 model of the same car, he told me that he had had the same problem with the rightside fan as I had with the '55 model. Thus, a problem on the '50 model was corrected in the '51, and through the evolution of change the fault was reintroduced on the '55 model. Evidently the company did not establish a standard as a result of their earlier experience. To sum up, to standardize a well-functioning design is to ensure that you will do no worse in the future. It does not mean that you do not try to do better when you think you have an idea worth exploring.

## Results to Be Expected

What can you expect to gain from the investment in standards? Dollarwise, at least six dollars average for every dollar spent. This is only in the direct measurable costs of parts, materials, inventory, procurement, etc., as reported by member companies of the American Standards Association. But what of savings not as easily measured?

What of the avoided arguments?

What of the material made right the first time?

What of the chemical baths and batches maintained correctly?

What of the simplified training?

What of the engineering time saved?

(My division saves over a million dollars a year by just having standards manuals available on designers' desks – this, with only about eight hundred manuals distributed.)

These latter elements are difficult to measure in dollars, but an astute manager should be able to grasp the full impact of smoothing up the operation – you can make enough in measurable savings to pay for the program, and a five-times return on top of that. Everything else is gravy.

**Other Benefits**

Designers need not dissipate their energies on "reinventing the wheel." That is, once they have worked out basic design approaches that are common to much of the work, they need not repeat this effort. Regardless of your line, it is probable that you can separate any job into those features which are unique and those which are repetitive. Thus, if you establish company standards to cover repetitive situations, each job consists of engineering its unique aspects and specifying the necessary standards which will tie the job together. If you are building machinery, then a new design would consist of engineering the special features of the machine and tying these together with your established standards on bearings, machine fits, joining hardware, raw materials, finishes, etc. If your business is building chemical plants, then you have already decided what pumps, piping, and fittings are most suitable for the usual line of your work, and thus your major task for each new job is to tie the standard parts together for the specific job. If you are designing automobiles, you may take advantage of the wide variety of SAE standards for parts, bumper heights, and wheel sizes, and use your design ingenuity to build power plants and to style and dimension the car in proportions that you think will sell.

Procurement expenses will go down for those categories of parts which are included in the standards program. You can expect this because you can make multiple use of a select family of standard parts; there are fewer parts to buy and these in larger quantities. Also, if the part selection enables you to establish a max/min inventory program, you can put the entire requisitioning process on somewhat of an automatic basis. This is a prime benefit in parts standardization.

You can expect a reduction in the variety of raw materials which must be handled. These benefits accrue in two ways. First, there may be considerable reduction in the variety of materials for the stockroom to handle, or enough volume usage of fewer materials may result to enable you to stock where you could not before. A secondary but nonetheless important benefit of raw material standardization is that it helps you to refer to materials, available under several trade names, by some common nomenclature provided by a national standardizing group or the military. Examples which come to mind are the variety of tool steels available from many steel companies, each under a trade name, but all of which comply with a single federal specification. Another example is the plastic laminates which are available from several laminate manufacturers, each under a different trade name, but are identified by uniform NEMA designations. More than one designer has called for a piece of material by trade name, only to be informed that it was not stocked, where in fact it was available under another trade name or under its generic description.

Sound company standards contribute in still other ways. They aid in predicting the reliability of your product, in terms of how well it will perform and over how long a period of time. They improve serviceability of the product as you learn how to build better for servicing. They enable you to predict costs to the point where you save money where you can afford to save it and spend money where you have to, knowing that you are doing these things correctly rather than by wild guesses.

Standards also enhance your ability to produce a second lot of equipment or parts or materials which is like the first batch. (How often have you run into cases where you built a fine working model and then found that because this model was hand-tailored and all requirements were not set down in the drawings or specifications, it was impossible to build a second like the first one without a lot of rework?)

### The Role of the Standards Engineer

To ensure success, a company standards program must be operated by someone or some group to whom the program is a prime assignment. Once the program becomes a secondary job, the demands of the primary occupation tend to preclude any real and continuing attention being given to standardization. The standards group must think standards and people and savings (and also about spiking the guns of tradition that are always ready to shoot down a new idea). Surely you have heard some of these replies to fresh proposals:

"We tried it twenty years ago and it didn't work."

"We've never done it that way before."

"A good idea but it belongs in another department."

"If it were any good, it would have been thought of before."

"We're not ready for it yet."

I grant you that some proposals deserve a quick death, but not all of them. The standards engineer has to be prepared for these reactions and to persevere, but to do it tactfully. He also has to be certain that what he is proposing does in fact meet the needs of the company. However, everyone sees a problem from his own vantage point – often a very narrow one. The standards engineer cannot afford such a luxury, nor can he afford a premature conclusion. Sometimes the last bits of information gathered during an investigation can alter its entire course.

The standards engineer has to be persistent, else he will be easily discouraged. He must carry the ball continually, otherwise the standards project will wither away. No one else should be more interested in seeing a project grow into an approved standard than the standardizer himself.

He must sell standardization as a technique and as a state of mind at every opportunity. Utopia would be a company that was so standards-conscious that everyone wanted a standard to cover all routine situations. Then every engineer and manager could devote most of his attention to the special problems that come up with each new project and each new day. We often spend so much time putting out daily "fires" that we have no time left to plan for fire prevention. This is a land of "millions for fire extinguishers but not one red cent for fire prevention," to paraphrase an American statesman of many years ago.

All of this means that the standards engineer cannot sit around and wait for project assignments to come to him. Others may suggest a project, but only the standards engineer can evaluate it in overall terms. Only he can decide whether it should be added to his agenda or not. True, a standards committee may make assignments, but only after close investigation, probably by the standards engineer himself, as to the potential.

**Publishing the Company Standards**

I don't want to give the idea that a company standards systems has to add several tons of paper to the files. Companies usually start their standards program with one loose-leaf binder subdivided into sections which are logical and germane to that company's affairs. For an equipment manufacturing firm that uses large numbers of mechanical and electrical parts, and, of course, requires product drawings, a first issue of the manual might consist of sections on design features, mechanical parts, electrical parts, and drawing practices. As their size increases, these subdivisions then lend themselves to establishment as separate manuals.

Subsequent to the initial program, other sections may be added on shop processes, raw materials, finishes, quality control, and reliability requirements. In short, sections or volumes dealing with any company function requiring agreement are in order. Thus, you start with the most urgent problems and expand only as required. In essence, each standard need be no more than a record of agreement between affected departments. This may be a set of minutes of a meeting or it may be the formal agreement resulting from a lengthy investigation by a standards engineer. Regardless, it always should result from a consensus among those having substantial interest in the project.

You will find that your various engineering or manufacturing specialists will benefit greatly from having the support of a few key standards in their field of interest. This has the effect of getting more people in the organization acquainted with the more common problems and their correct solution. This relieves your specialist of the burden of continually guiding and reguiding people in the most rudimentary aspects of his specialty; he has more time to concentrate on the more difficult and unique aspects of each design. Of course, there is often the strategic problem of initially convincing the specialist that a standard in his field will not necessarily put him out of a job. After he gets his feet wet and realizes the full benefits of writing standards, you will probably find that he is an ardent supporter of the standards program.

### What Standardization Is Not

*Standardization is not a straightjacket* which restrains your movement once you have it written up. Dynamic standardization calls for standards to be modified, updated, improved, challenged, and in every way kept valid for the problems of the day and not permitted to become archaic solutions to yesterday's problems. (In fact, this necessity to maintain standards current plagues every standards activity in that every new standard generated is one more to maintain; manpower must be devoted to an ever increasing volume of maintenance work, thus reducing the ability to work on new projects.)

Standardization is not the dusty image that many people think. (In fact, if the standards manual is so seldom used that dust is allowed to collect on it, it should be withdrawn from the manual holder and given to someone who can make better use of it.) Company standardization is not merely a dull, drab exposition of table after table of technical data, although many standards may in part consist of just this type of information. Rather, a company standard is a document recording agreement among progressive-minded people, showing that they are interested enough in a common problem to spend some time giving it their considered thought. A standard must be dynamic rather than static, lest it lose its usefulness quickly.

*Company standardization is not necessarily a series of mandatory edicts* issued by top management without regard to specific applications. This is not to say that some select group of topics should not be mandatory, such as safety standards and other standards affecting the welfare and well-being of personnel, or the methods of handling payments of large checks. However, in the usual type of engineering standard, the company standard should not be mandatory in a given situation but rather the consideration of its use should be mandatory. Only in this way can final control of product configuration remain in the hands of the designer. He is the only one in a position to know all the detailed requirements. Nonetheless, he should be required to consider the use of any existing company standard dealing with the topic area involved; and if the standard is not suitable, this fact should be communicated to the standards organization in order than any necessary modification may be made, or a new standard generated. With this continual feedback from the people actually involved in design or production of the company's products, standards can be continually improved and kept current.

**Does Your Company Need Standards?**

Those who do not currently have a company standards program may be wondering how to relate all of these thoughts to your own company's situation. You would like to be able to decide that you do or do not need standards, and, if you do, determine how you can justify the expense of generating them. In answer, perhaps the most certain and pragmatic means is to check on the kinds of problems that cross your desk during the next week. Are they similar to the problems that you were called upon to decide last week or last month?

If you put a new man to work next week, will you have a firm training program in mind and can you direct him to a body of company-accepted information that he should master while working on his first assignment?

Walk over to your inventory files and scan the cards. Do you see material varieties and sizes listed that seem close enough that one should do the work of two or three? Do you find that you are stocking and handling machine screws in three or four head styles and in several materials and finishes?

Check your housekeeping supplies. Are you buying to satisfy the personal preferences of individuals rather than selecting a few staple items and buying in quantity?

If your company maintains repair centers, check over the items which they have to carry in stock to see how many similar items there are. This will tell you whether parts standardization could smooth out your service operation.

I can think of no more direct approaches than these to determine your needs. Another approach would be to have a chat with the purchasing agent to find out where he is directing the bulk of his purchasing effort. Is he spending large amounts of time buying low-priced, high-consumption items? Is he buying small quantities of slightly different items on a frequent basis? If so, engineering is limiting his ability to buy economically.

These suggestions will by no means constitute a detailed economic evaluation of the potential of standardization for you. However, they will serve to point up some of the more glaring areas that will benefit from a standardizing effort.

### Summing Up

There are three ways you can meet price competition today. First, you can cut your price to meet your competitor's or go below it. This could put you both out of business. Second, you can leave your price higher than your competitor's and hope that you will survive on reduced sales until he puts himself out of business. If, however, he has in fact reduced his costs to allow him to reduce prices, you then have only the third choice – you must cut your own cost yet maintain quality. A progressive standards program, actively supported by management and administered by an alert self-starting standards engineer capable of taking hold of a project and seeing it through to a logical conclusion, can be a major contributor both to reducing your costs and improving the quality of your product. Can you ignore this potential?

# THE INFLUENCE OF A COMPANY STANDARDIZATION POLICY

Edgar S. Kirk

Our vantage point is that of a company which has been going through the long night of labor pains to the birth pangs of a formal standards program. Our anticipation is high as we view the potential influence of such a program upon our growth. To be sure, we are well aware of the practical limits of a standards program. We know that any program which espouses a philosophy of standards for standards sake must be unsound and economically disastrous. However, we firmly believe that our standards program will greatly and beneficially influence every area of our company, directly or indirectly.

In the beginning, we attempted to launch out into the deep immediately after getting our feet wet. However, we soon found ourselves altering our course because of pressing demands upon engineering time. Slowly, but surely, we are getting back on course and hope to be under way at flank speed soon.

We believe the most influential way to introduce a standards program is to start with a clear definition of standards, and a declaration of purpose. What is wanted is a single system which will maintain and improve reliability and, at the same time, increase efficiency and productivity, thus decreasing cost. At first, this may sound like idealistic thinking; yet a closer examination reveals its practicality. All too often, systems which attempt to cure some particular disease soon degenerate and cannot be distinguished from the disease, or become a part of the disease. P. A. Sorokin, of Harvard University, said the curse of our degenerating culture is mediocrity, and this is the result of a lack of purpose and true objectives. Thinking which lacks true objectives, whether cultural or industrial, soon produces stagnancy and outright sterility. Rather than enhancing growth, it results in death because it suffocates in its own inefficiency.

The President of the United States says he is not as concerned about the growth of our gross national product as he is concerned about its quality. It is catastrophic to sacrifice quality for quantity, whether in the defense of our nation, or in the production of consumer products.

Through the maze which you will explore in a search for a practical standards program, there are two things which will make the greatest demands upon your time, talent, and energy. These are reliability and cost – the inseparable twins.

The only practical way to lower cost in the long run is through reliability. Harry Chesebrough, Vice President and Director of Quality Control and Reliability of Chrysler Corporation, said, "It isn't quality that is costly; it is the lack of quality which unbalances the books. Consistent low quality can mean complete loss of business, and that's as costly as you can get." Therefore, we reiterate, reliability should be the prime motive in a search for a practical standards program which is to influence every phase of the company and generate new growth.

Without this motivation and practical objective, a standards program will echo the lines of Macbeth, ". . . a tale told by an idiot, full of sound and fury, signifying nothing." Standards must be looked upon as a means, not an end.

Let me give you an example, from our own company. Bostitch, like other companies, has worked to many standards throughout its long history. (If it hadn't, it wouldn't be here today.) However, these standards were never systematized and properly documented for the most efficient and profitable use. The result: duplication of effort, duplication of standard parts, and loss of engineering time because of inefficiency in information retrieval.

Out of the sudden pressures of World War II, and its demands upon industry, came the realization of our growing need for a formal standards program. Token programs filled the gap, but they did not meet the need. No real action was taken toward a formal program until a few years ago, when one began in the product engineering department.

It is interesting to note that this program was greatly influenced by an invitation from the Rhode Island Section of the Standards Engineers Society to attend their Annual New England Regional Meeting. The interest and enthusiasm generated at this meeting put us on the path of formulating a practical standards program. We are grateful to the Society for the assistance and encouragement we received from its members, and the valuable professional advice so freely offered.

### Suggestions – Formulating a Practical Standards Program

To be sure that your program will be the most influential and generate the greatest growth, make a careful survey of your needs. We believe they will be:

1. The need for a common language. Investigation will reveal the fact that most recurring problems are related to semantics, not to engineering or design. Effective communication in the exchange of ideas and in the dissemination of engineering specifications and requirements must be based on a common language.

Let me give you a prize example of something which could have been avoided through more efficient communication. Our representative in Australia ordered six complete clutch assemblies for one of our stitchers. The foreign order department sent the order to product engineering with the request that we list the parts required, because they are not stocked as a complete unit. One of the parts of this assembly is a large, weighty cast-iron flywheel. A few days later we received a cable from our representative urgently requesting that we ship airfreight as soon as possible. The wailing which occurred upon arrival could be heard clear back to the factory. Australia did not want the clutch with the cast-iron flywheel and they put in a claim for $277.00 "excess" shipping costs. All of this could have been avoided through proper communication.

2. The need to eliminate repetition. Not only is this costly in terms of wasted effort, but it often creates new problems and adds to the existing confusion.
3. The need to reduce time spent in nonproductive areas.
4. The need for more effective information retrieval.

After making your initial survey, establish some immediate goals and then initiate a basic program. But the age-old question will arise, where is the best place to start? There is no pat answer: this will depend upon your individual need, which may be drafting standards, design standards, procedures, material standards, or manufacturing standards. To get the most out of your program, pick the area of greatest need. The valuable influence of standards on cost reduction, whether in terms of parts or effort, will soon be evident to all.

Let me give you an example. If your immediate need is to catalog standard parts, there are two quick ways of doing it. If you have a data processing system, it is a simple matter to get a numerical list of all active standard parts. From this list the catalog can readily become a reality. If you don't have a data processing system, run through the stock cards and list all active standard parts. From this list the catalog can easily be made. Such catalogs greatly reduce costly engineering time wasted trying to find required components. Up-to-date catalogs also deter men from becoming disgusted and ordering duplicate parts, increasing inventory, and adding to the existing confusion.

To give you some idea of how influential we hope our standards program will be, allow me to list some of our working goals. These were established for design standards, drafting standards and procedures, and material standards. They are as follows.

A. Objectives of Design Standards

1. To provide more time for creative effort through documentation of classified information covering available materials and components
2. To simplify product design and reduce costly complicated manufacturing
3. To reduce product variety and eliminate repetition
4. To establish permissible alternate designs
5. To furnish technical data and availability on recommended materials, thus minimizing use of costly exotic materials requiring special processing
6. To furnish documented design and technical data permitting shortened development and preproduction time for new products
7. To avoid extravagance through practical product and quality requirements, maintaining required quality for highest product performance and reliability
8. To establish practical tolerances and allowances

B. Objectives of Drafting Standards

1. To establish efficient communication between design and manufacturing
2. To provide for effective control and dissemination of product engineering specifications and requirements

3. To maintain uniformity in graphic communicaton through
   a. Clarity of details
   b. Uniformity of nomenclature and terminology
   c. Consistency in specifications furnished to vendors
   d. Uniform format for detail and assembly drawings
4. To insure 100% interchangeability of components
5. To establish and maintain quality and reliability through
   a. The use of accepted standards
   b. The use of technical standards
   c. The use of American standards
   d. The establishment of new standards soundly based on product engineering criteria and product performance requirements
   e. The establishment of testing procedures to maintain optimum efficiency and performance of products
6. To establish procedures for effecting design changes at minimum costs
7. To establish criteria and format for technical literature – such as customers' instruction books and parts charts, and service personnel machine service manuals

C. Objectives of Material Standards
1. To furnish designers with available and recommended materials
2. To furnish metallurgical data on materials in steel stores, including machinability, weldability, and hardenability
3. To minimize inventories of extravagant exotic materials
4. To eliminate duplication of commerical components and fasteners

To be certain of our growth and to prevent wasted standards efforts, our standards are mandatory; and deviation from these standards on engineering documents must be reviewed and approved by the manager of product engineering.

All of our standards are based on recognized national standards, technical developments, standards of trade association, and previously established performance requirements of our products.

At this stage in the evolution of our standards program it is difficult to know how influential it has been in the growth of our company. However, we are confident that it will be influential because motivations are sound and the program is based on time-tested and proven principles. We believe our future is bright and our growth is assured because we are helping our people do their job better and more efficiently through the practical application of standards.

# MEASURING YOUR STANDARDS DEPARTMENT TO A STANDARD

John L. Rayfield

The words "standards" and "standardization" are two of the most misused and misunderstood words in the engineering profession. As a result, the "standards department" is the most misunderstood department in the company. In those companies that have standards departments, I dare say that very few people know what the standards department does, or what it should do. Many people think of standards as "screws," "nuts," "bolts," etc. – a far cry from standards as we know and can practice them today.

I remember reading that at an investigation of high military costs, a layman said he did not see why the government had to pay $5.00 for a gross of screws made to a military standard, when the same size and type of screw could be obtained for $1.00 in the neighborhood hardware store. While we who are in the business know that these two screws are as different as night and day, he was unconsciously making a gross error due to his lack of knowledge.

## Shall We Then Change Our Name?

Because of such misconceptions, many standards departments are attempting to change their image by changing their name to one that is more indicative of their time, status, function, and responsibilities. After much consideration and deliberation, however, very few have come up with a suitable alternate name. Does this indicate that either our responsibilities are so vast that "one word cannot indicate the true overall standards function," or that we do not know what we are doing or should be doing? In either case, it is time for us to review our position and what should be done.

We might start by saying that things equal to the same thing are equal to each other; and equate the two previous statements by saying that we are talking about the same thing but really stating it in two different ways. Thus, we cannot run from our problem, but must meet it head on, resolve it, and go on to the next problem. This is the problem for today, "How do we enhance the image of the standards department?"

I do not believe that there is any new or different single word that describes the standards department's activities better than the one we now have. I would recommend, rather than changing the name, that we keep the word "standards" and educate and promote standards activities.

### The Question Is, How?

Perhaps the best way is to create a proper image which is understood and is indicative of the standards department's overall activities. This image should present the overall and detailed responsibilities, breakdown of various areas, organization, and any other important factors.

To create this image, I am recommending a standard for a standards department – a standard to which you can compare your present responsibilities, a standard to which you can measure your present scope, your present organization, your present effectiveness.

It is said that standards engineers are the most unstandardized of engineers. This is probably true; however, the standard I am recommending will make us the most standardized and organized. People will know what standards stand for in their entirety, and we will have a means to present management with a program for an effective standards department, all-inclusive, definitive, and the same throughout industry. For the first time we will have a valid image – which will be the same regardless of whom we employ or where they came from. Then and then only will we have achieved the true image of a standards department, one which is standardized and is measured against a standard, one which management will come to recognize as important and needed.

To those companies that do not have a standards department, it is time that they did, time that they recognize the importance and necessity for creating one. They may not realize it, but they are actually doing standards work, but doing it in a nonprofessional, not too productive manner; probably it is being done as an odd job by technicians, draftsmen, designers, and others not trained for the job. Invariably they will be doing redundant work, following unorganized, improper procedures and communication. All this adds up to a waste of time, money, and manpower. To these unstandardized groups, I recommend that you make your management aware of the benefits of a true and competent standards department.

The standard I am recommending is shown in figure 1. The standards department is divided into two categories, "Data Services" and "Technical Services." The Data Services group is subdivided into three areas – Standards Library, Standards Manual, and Checking (Data & Documentation). The Technical Services group is subdivided into four areas – Component Selection and Application (Quality & Reliability), Materials and Processes and their Selection and Application, Specifications for Equipment and Systems, and Standards Laboratory for Test and Evaluation. Figure 1 also indicates those items for which each group is responsible or to which it gives support.

To a few of you, figure 1 may seem commonplace and quite acceptable; to others it may seem too complex and not necessarily adaptable to your company or work. However, to some of you it may mean nothing. This is the group in particular which this standard will most benefit. Persons in this group have accepted their position, their organization, their place in the company. They have lost their capacity to improve, to change, to accept new ideas and concepts, to plan and look forward. They must change their point of view and image from the negative and passive to the positive and forward moving.

The effectiveness of the standards department depends on where it is situated in the company organization chart and to whom it reports. The position it should achieve is one in which it can be most helpful without being too far down on the organization chart. The ideal position is shown in figure 2, where it is part of a group known as "Engineering Product Assurance." This in turn is headed up by a chief engineer reporting directly to the vice-president of engineering or his equivalent.

The important thing is that the standards department must be at least on the same level as, or higher than, the engineering project managers, chief draftsman, chief designer, planning director, program manager, and reliability and maintainability manager. When the standards manager is at the proper level, he can participate fully in meetings, discussions, programs, proposals, contracts, and, therefore, can be the most effective for his company. It is interesting to note that many companies are

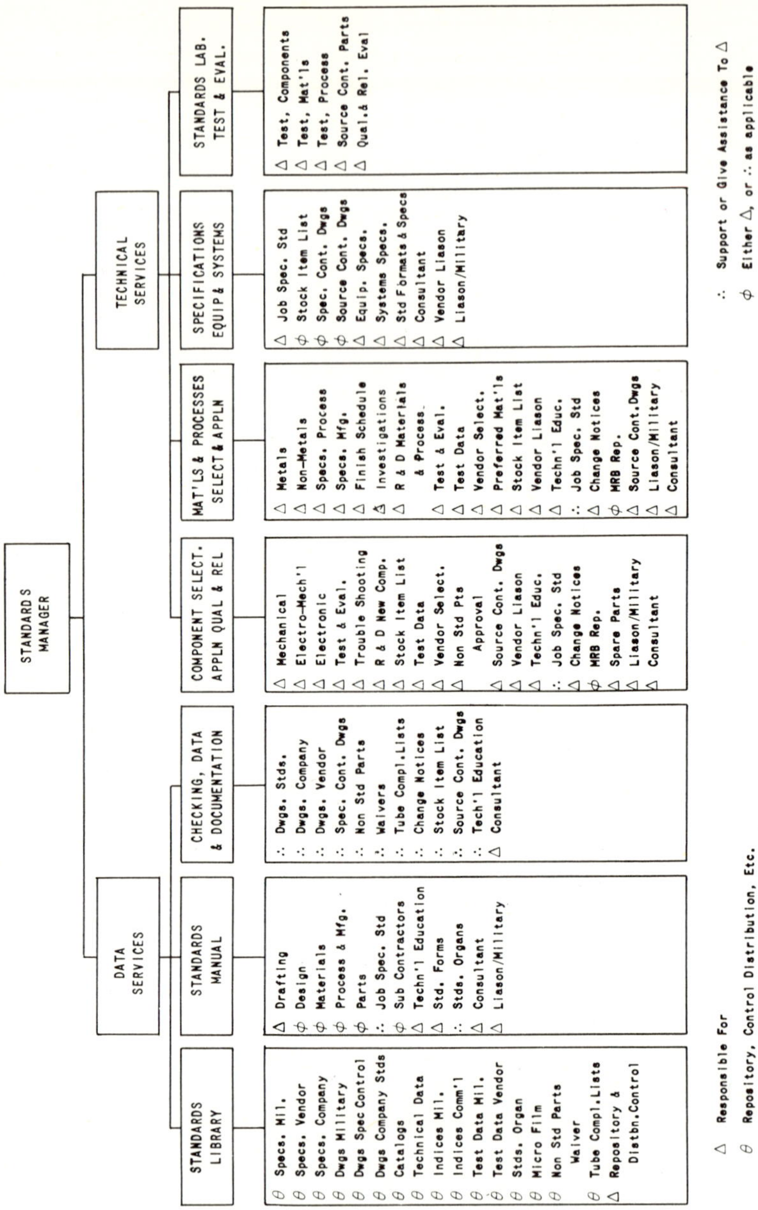


**Figure 1.** Standards Department Organizational Standard.

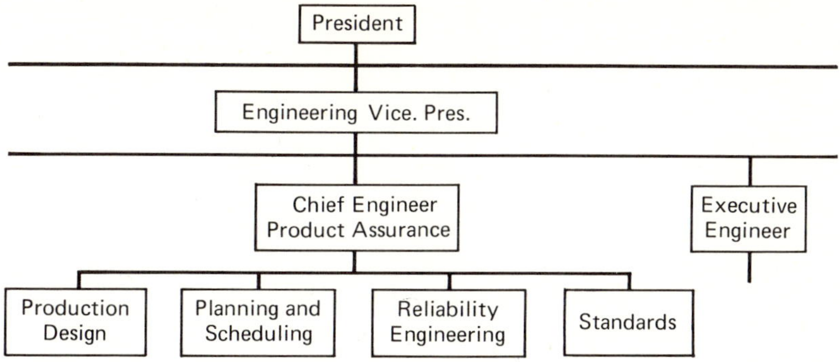


**Figure 2.** Standards Department in the Company Organization.

reorganizing their engineering departments to improve their engineering capability by having a parallel "Quality Assurance" program in engineering similar to the "Company Quality Assurance" program in manufacturing. This would be in keeping with the new "Zero Defect Program" now being instituted and promoted by the Department of Defense. Should your company have an "Engineering Product Assurance" Department or its equivalent, that is the recommended place for the standards department.

## Conclusion

May I reemphasize the importance of measuring your standards department to my standard as an aid to progress. With your help, initiative, and interest in doing a good job you can improve your company by using this standard to build an effective standards department. Last, but not least, you should use this standard to raise the image of the standards profession. I urge you to establish and promote the standards department as a functioning and recognized group with proper status, responsibilities, support, and position in the company and on the organization chart, at a level which will not hamper your operation and which will be effective when directives are issued.

# Part 4

# BENEFITS OF STANDARDIZATION

*We showed that a good standardization program is a program for good management. Therefore, where there is standardization there are always cost savings. Good management is always economical.*

*A great number of articles appeared on this subject. Too often standards are sold as a cost reduction means, but emphasis on only the cost reduction aspects of standardization may hurt the cause of standardization*: It obscures the real purpose of company standards, which is not cost reduction but greater profits.

*In this section we will see many examples of how standardization reduces cost. All that the authors said here is absolutely true; however, it would be useful for you to know one man's opinion about the relationship between standards and cost reduction, something I wrote in April, 1964, issue of* Standards Engineering:

> *Investment: $6.00 for a Dollar! This is such a wonderful return on an investment that it reminded me of a time when, many years ago, somebody wanted to sell me the Brooklyn Bridge. It is not that I doubt for a moment that Mr. XXX has achieved handsome dollar savings for every dollar invested, but to make a general rule out of this and consider it as a standard return on any dollar invested in standardization is rather dangerous self-kidding.*

*Saving is done by reducing waste. The greater the waste, the greater the savings that can be realized. Could you save $4,000,000 in a company which has $4,000,000 or less business? Let's take a hypothetical case:*

| | *Company X* | *Company Y* |
|---|---|---|
| *A company making* | *$1,000,000,000/yr.* | *$100,000,000/yr.* |
| *has stock of* | *$10,000,000* | *$1,000,000* |
| *hired to reduce stock* | *5 Standards Engrs* | *2 Standards Engrs* |
| *Standards Engineer's salary (theoretical)* | *$50,000* | *$20,000* |

*Stock in both cases was reduced by 10%, but return on investment was as follows:*

*Company X: $1,000,000/$50,000 – $20/dollar spent.*
*Company Y: $100,000/$20,000 – $5/dollar spent.*

*During one year, standards at a great company were used by 6,000 engineers over a million times. At the same rate, a company with only 600 engineers will use the standards only a hundred thousand times. But the expense to create such good standards as that of Company X will be the same in time and money spent regardless of the number of people who will use them!*

*About cost reduction: We must understand that the main purpose of standardization is not cost reduction, but profit. This is achieved by optimum utilization of all company resources – men, money, materials, processes, equipment and time – to widen the company's profit margins. And this objective is achieved by means of written managerial decisions (standards), which insure a technologically and economically consistent product.*

*Standards engineers must be identified with profits, and not with cost reduction; with management, and not with rank and file engineering. A good management always has cost reduction as a sideline. Cost reduction alone does not necessarily mean profit or even good management, and could be just an effort to reduce losses, to the actual detriment of company operations ("penny-wise, pound-foolish"). Profit is something completely different – in order to assure profit, cost must often be increased.*

*If cost reduction was the sole purpose and justification of standardization, as ASA and too many members of SES publicly attest, then we should call ourselves value analysis engineers and forget all about standardization. Actually we used the VA method a long time before the term was invented. Any time we standardize, we must value analyze, but our function is much wider.*

*Anyway, whatever the nature of standardization is, the fact still remains that innumerable surveys and the experience of all standards engineers proved without a doubt: Well organized standards departments do bring considerable savings.*

*Different advantages which result from standardization are covered in this section.*

*R.G.*

# THE BENEFIT OF STANDARDIZATION TO INDUSTRY

Leslie D. Price

## The Question

Standards! What do standards mean to industry? Efficiency, economy, and effective management.

Broadly, we may consider standards in industry as purchase specifications, manufacturing standards, materials specifications, finished product standards, test methods, safety standards, and contract conditions of sale, among others. But what, basically, makes standards of primary importance to industry? The answer may be summarized in one word: economics.

We, in the United States, are today faced with a basic struggle for survival: survival of a way of life, survival of life itself, if you will. Call it the "cold war" or any other name you may wish to use, but the fact remains that the "heat is on." We are locked in a life-or-death struggle with communistic or sociological ideologies. The battle ground may be military or it may be economic. At present, it is largely economic, with the well-grounded fears of extermination through military struggle. Aside from the political implications of the problem of the world struggle for survival, the principal factor is economic leadership or advantage.

A prominent factor in maintaining our economic position involves competition in world markets, including our domestic market, with our allies, to whom we have given or loaned a great deal of money and know-how to reconstruct their industrial economy following World War II. Our friends are substantial competitors due to lower wage rates, a willingness of foreign labor to deliver a full day's work and output, use of modern methods and tools without featherbedding, and the willingness of governments to subsidize exports in order that necessary dollars, or "hard money," may flow into their countries.

International standards enter strongly into this picture. If products which meet American standards do not conform to international standards that are being promoted for worldwide use, we are substantially out of the market. With standards set, the market is virtually sewed up.

One of the most important factors in industrial economics is standardization. It is a most important tool of industrial management, and today I should like to look at standardization with you from the eyes of management and what it means to the economics of industry. I believe that, with such a look, we shall see "The Benefit of Standardization to Industry."

## Industry-Labor Responsibilities

If we are to keep our free enterprise system, industry and labor must both recognize their basic responsibilities. To the public, industry must supply an ever-increasing amount of goods and services at reasonable cost. Labor looks to industry to provide more and steadier jobs and to increase, as far as practicable, the real income of workers. Industry can only accomplish this if labor also accepts its responsibilities.

### *Management Obligation*

Management has an obligation to utilize to its utmost those factors and procedures –

1. which will permit creative effort to be used to its maximum efficiency
2. which will use human and material resources to best advantage
3. which will create better understanding and coordination between industry, its customers, the public, and government. To accomplish this, management must plan, organize, integrate and determine accomplishment for the maximum benefit to the company and to the public. Standards give management a rifle shot at its primary targets of planning and evaluating results. Without standards, a wild shotgun aimed haphazardly at the target might by luck score a bullseye, but this is certainly no scientific approach. Without standards, management can become tied in the red tape of detail, inefficiency, confusion, and general malfunctioning of the industrial process.

### *Labor Obligation*

If labor is to retain its freedom and strengthen its economic position, it has an obligation to recognize the need for increased productivity.

Labor-saving devices, standardized methods and procedures, and criteria of efficient performance are tools which labor should employ to meet this obligation. Important to increased efficiency and productivity is the avoidance of accidents. Labor has an obligation to itself and to its employers to use to the maximum the safety standards established to minimize hazards.

### What Management Expects of Standards

Management uses standards to help crystallize knowledge in the direction of greatest benefit to industry.

Research and development, design, manufacture, sales, public relations, and the use of products need constant evaluation. Management looks to standards for relief from executive decisions on those matters which are already documented.

The use of standards in this country is entirely voluntary, and it must remain so. It is, therefore, incumbent on management to recognize the powerful potential of standards as a tool, and to let its employees know that it uses and expects the use of standards; otherwise, employees are apt to do nothing about their use.

### National, Industry, and Company Standards

Before proceeding further, we should review briefly what we mean by industrial standards.

Standardization is an extremely broad term. Industrial standardization may be considered as falling into three general classifications: national, industry, and company. While many standardization activities are so closely interwoven and interdependent that it is often difficult to draw a distinct line of demarcation between these three classifications, they may be defined as follows.

*American standards* are national in scope and are developed through the all-party-at-interest procedure of the American Standards Association and hence are generally recognized and accepted by producers, consumers, and general interest groups.

*Industry standards* may also be national in scope. They are considered, broadly, as those developed by industry trade associations, professional engineering societies, governmental bodies, or other associations or engineering groups.

*Company standards*, as the name implies, are confined to standards and specifications developed for use within a company.

Basically, the same fundamental standards obtain at all levels. Obviously, they must be consistent. The industry standard is generally more detailed and covers phases of the subject which may be neither necessary nor desirable in the American standard. Similarly, the company standard is basically the same as the American and industry standard but covers the subject in considerably greater detail than either of the others.

Company standards and specifications go considerably further into the details of construction. For example, the specifications for steel and iron for the manufacture of the rotors, frames, and shafts of motors could not be included in industry standards without tending to stifle design and progress. The company standards may include details of composition and characteristics of all raw materials entering into the manufactured product and generally form the basis for purchase specifications. For these purposes, specifications and standards of such organizations as ASTM are utilized freely. Dimensional tolerances of component parts must be specified in order to make assembly possible. Testing standards are included all along the production line in order that the final assembly may meet the performance specifications of industry and American standards. I like to look at this concept as illustrated by two triangles or pyramids, one inverted as in figure 1.

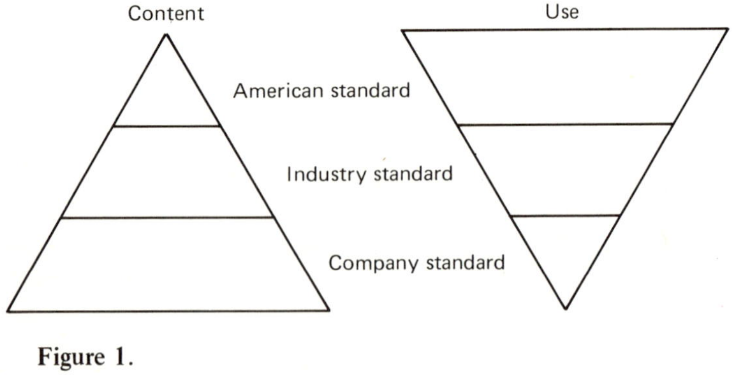


**Figure 1.**

The first pyramid illustrates the broad use of content of the company standard diminishing to the American standard, and the other the broad base of usage of the American standard diminishing to the company standard.

Many of us accept industry and national standards as a matter of course. If questioned, I think everyone would admit that standards are indispensable tools to our industrial economy. Yet, probably relatively few of us have analyzed why. Standards are used daily without thought or realization that we are using them, almost as we use the other common tools of industry without thought. Little do we stop to think of the utterly hopeless situation which would result in the manufacturing world without standards. Having them, let us see, through the eyes of management, what is expected of them and what they actually do for us in making so much available for so little.

### Research and Development

Research provides oxygen to the fire of industry progress.

Research and standards appear at first glance as incompatible. They are not. To explore the role of standards in the field of research and development, we must first define the areas we are discussing. Research, as generally applied in industry, may be divided broadly into three areas:

1. Improvement of present products.
2. Development of new products, processes, and uses.
3. Fundamental research – original exploration into "the wild blue yonder" without special consideration of existing products or products contemplated in the immediate future. In other words, it is basic scientific research, the results of which may become useful in the present field of a company's activity, or some other field. It may contribute to the whole scientific knowledge without commercial significance to any one industry.

What does management want to know about research in any of these areas? Only this – management is responsible for financing such research and for evaluating the results of research in terms of immediate or ultimate financial return to the company. No manufacturing company can afford to be an eleemosynary institution. Many results of research benefit the great store of technical knowledge generally. However, we cannot lose sight of the fact that the real motive for industrial research is profit. It is to learn new facts which will give a company a stronger marketing position by developing new, better, or less expensive products than those of its competitors.

Management must evaluate its own research and development programs. Industry standardization enters into this in two ways – knowledge of existing standards which establish platforms from which development may spring, and knowledge of what is being done or is being considered by others in the same field. The latter entails active participation in the standards activities of trade associations and technical societies related to the product or products under consideration. Existing standards establish a bench mark for comparison of research results. What is contemplated in the standards field provides information of the thinking and the development of others. Without this knowledge, valuable time and effort may be wasted in duplication of effort. Research, in turn, elevates the platforms periodically to higher levels of performance and perfection of products.

In research and development, the utilization of existing standards for components, materials of construction, and the present use of products may be quite helpful in evaluating new developments.

Standards of nomenclature, symbols, parts designation, etc, are essential to a common understanding of researcher and management. The common language derived from standards is necessary for communication. The research department may have "the development of the century," but if management doesn't understand it, it is wasted time.

As automation is applied more generally to industrial processes, and as the rate of change in technology accelerates, it becomes more and more necessary to recognize those problems which have been solved and to leave the research scientist free to develop.

### Standards in Design

Continuing into the fields of design, manufacture, quality control, and marketing, economics is the prime consideration of management. What does management see in the full utilization of standards in those areas?

The use of standards simplifies design problems. Two of the basic considerations which the designing engineer must take into account are performance and dimension.

Performance may include stress, strain, rigidity and other strength and mechanical considerations, or it may involve the operating characteristics of components, such as those of motors and control equipment. Design coordination of the characteristics of materials and components is always an essential for a properly functioning product.

The preparation of specifications for materials and components used in the construction of the device is materially simplified through the application of national and industry standards and specifications, such as those of ASTM and NEMA.

Performance standards facilitate the selection of components to achieve the desired results. For example, in the design of a piece of equipment operated by an electric motor, the operating characteristics of the completely assembled machine determine the necessary starting torque, the speed, whether continuous or short-time duty is required, the service conditions under which the motor will operate, the type of enclosure necessary, and the rating. An electric motor may be selected on the basis of established performance standards with the assurance that the motor will perform satisfactorily.

Recognizing the necessity for interchangeability of parts, dimensional standards facilitate design procedure. The use of standards in the design of components of assemblies not only facilitates the location of the components with respect to each other, but also assures complete interchangeability of parts, not only from a dimensional, but from a functional point of view as well.

Again using the motor as an example, dimensional standards such as those contained in the NEMA motor and generator standards assist the designer in selecting a motor which will fit into the assembly physically. Frame sizes, frame mounting positions, and flanges give a wide range for selection of motors for a given application.

Selection of the method of mechanical drive is facilitated by reference to the motor standards on mechanical drives. Flat belt, V-belt, gear, chain, or coupling drive standards give both performance characteristics, such as speed and horsepower limitations and dimension limitations of pulleys, sheaves, and sprockets.

By eliminating or minimizing misunderstandings in the specifications of materials, parts, dimensions, performance characteristics, and ratings, standards of nomenclature and dimensions further facilitate the design problem. Through such standards, the designing engineer knows what to expect in the line of performance, dimensional tolerances, and other characteristics. Likewise, the manufacturer or supplier understands precisely what the designing engineer is specifying. In other words, the purchaser and the supplier are talking the same language. Misunderstandings are avoided and costly errors are obviated.

Where standardized materials and components can be used, the work of preparation of specifications by the designer is minor. Consider, for example, the specification of a 5-hp, 220-volt, 3-phase, 60-cycle, 1,800 rpm squirrel cage, general purpose motor with class A insulation, class B design, open type, having 254 frame size, horizontal mounting with feet, and with standard terminal location. This statement is simple compared with the pages of performance design and dimensional specification which would be necessary without the available motor standards. Yet this specification tells the complete story, understandable to all suppliers.

Management sees the results of such use of standards far beyond the area of design. Hours of work are eliminated in –

Preparation of detailed specifications

Preparation of drawings

Research and laboratory test to determine performance compliance with specifications

Preparation of detailed purchase specifications by purchasing agents

Supervision

In addition, the work of clerks and typists is minimized and the files are not overloaded with voluminous drawings and specifications.

In the manufacturing process, the use of standard materials and components results in:

Fewer items to be purchased with correspondingly fewer orders, fewer ordering specifications

Lower storage and issue costs of materials and components

Fewer stock bins and less storage space

Fewer stock cards and filing space required

Fewer clerical recordings

## Standards Reduce Production Costs

When a manufacturer produces standard products, the total production as the result of the use of standards is spread over fewer items, thus providing greater production of each individual item. Standardization, of course, demonstrates its greatest economic worth in the manufacture of items that are produced repetitively.

In the ultimate, the repetitive manufacture of standardized equipment or material eliminates changes in assembly lines, thereby increasing the rate of production. As deviations from the standard are introduced, interruptions in repetitive manufacture occur and the rate of production of any one item decreases. Fewer changes in production, of course, reduce the machine set-up time, supervision, inspection costs, and also the probability of error.

Of utmost importance in the manufacturing process is the matter of maintenance and repair of equipment. Standardization of parts by the manufacturer saves both the user of the equipment and the producer. Standardized parts obviate the necessity for the replacement of defective or worn-out parts by duplication through specification of the part used in the original construction. A standard part can be ordered and the replacement effected. This obviously eliminates lengthy and detailed specification for replacement parts from the user of the equipment. Likewise eliminated is lengthy correspondence between the user and the supplier. Without the use of standard parts, voluminous blueprints and files of individual designs and specifications would be necessary to coordinate the specific design with each type of assembly. By the specification and use of standard parts throughout, insofar as it is practicable to do so, coordination of design of production facilities and maintenance is effected.

The use of specifications deviating from nationally recognized standards oftentimes necessitates the use of material not meeting the specifications and which would normally be rejected because of the necessity of meeting production schedules. This is, of course, unsound operating practice but is sometimes necessary.

The use of standards in production saves engineering time, clerical time, and purchasing department time and assures a better relationship between producer and user.

### Production

Individual pieces of equipment used in the manufacturing plant are not isolated units in the production process. They are, rather, integral parts of the production system. If a unit fails, the whole system is out of operation.

*Maintenance* is always important. Production line shutdown is costly. Partial or complete restoration of operating facilities is speeded up by the use of standard components in the system. The use of standard components in the design of production equipment simplifies and reduces the stock of maintenance equipment and materials, together with the necessary records and the procurement of replacement parts. The use of standard equipment also minimizes the necessity for makeshift repairs.

*Safety* is of primary consideration to every employer in promoting the health and happiness of his employees. Aside from the human aspects of management's interest in safety, accidents are costly.

Safety standards help protect the life and limb of workers in many phases of manufacture.

# COMPANIES REPLY TO STANDARDS SURVEY

from ASA Survey (Presently USASI)

Do company standards programs pay? Men who work with standards know they do, but evidence to support their knowledge has always been uncomfortably meager.

Now, results of a survey conducted this year have been released by the American Standards Association. In its role as the main source of information on national and international standards in the United States, ASA asked its 2,000 company members and about 800 nonmembers for information about their experience with standards. ASA wanted to know how many companies have formal standards programs, how their standards work is organized, how standardization is financed, how standards are used, whether the company cooperates in standards activities outside the company, and whether there are records of savings through standardization.

Questionnaires were sent to about 2,800 companies, and 238 were returned. Of these, 89 reported a formal, organized program. Thirty-four of the 89 were able to report actual dollar savings. Others, however, stated or implied savings in a general way, and several indicated that company policy prevented them from revealing dollar figures. A number indicated that intangible savings probably exceeded the tangible figures available.

Of the 238 who replied, 181 were members of the American Standards Association.

The companies answering the survey ranged in size from a two-man, part-time technical writing firm to corporations with several billion dollars worth of sales. By industry, the companies indicating standards activity fell into the following industry classifications, listed in order of number of companies included: Electrical manufacturing, general machinery, electrical utility, machine tools, chemical, gas utility, steel and iron, electronics and communications, petroleum valves and fittings, fasteners, aircraft, building materials, bearings, engineering and construction, railroads, instruments and controls, copper and brass, office equipment, photography.

The result indicates that those companies having formal standards programs, with budgetary control of standards work, are better able to show clearly the monetary benefit of standardization than those whose standards work is a part of other functions and under the control of a number of different departments. Although a total of 209 companies indicated either specifically or in a general way that they engage in some form of standards activity, the survey seemed to indicate that the majority of companies are missing the benefits of a standards program and of budgetary control of standards.

The most conclusive finding, the report indicates, is that American industry as a whole probably doesn't know the answers to questions about savings due to standards. Compared with information previously available to ASA, the report indicates, however, that more and more companies are beginning to realize the cost-cutting potential of a standards program.

Several of the companies reporting indicated that, as their new standards programs get underway and as results begin to be felt, they will find ways of pinning down the dollars-and-cents results.

One of the significant findings, ASA believes, is that 30 of the 34 companies reporting dollar savings also report extensive use of standards and standards manuals. This seems to indicate a close link between organized standards work and recordable cost savings.

Savings reported by the 34 companies that were able to quote specific figures were as high as \$50 saved per \$1 spent on standards work.

In terms of sales, savings ranged from 0.3 percent to 5 percent of sales, with an average close to 1 percent. In the introduction, the report calls attention to the fact that, on the basis of a gross national product of more than \$450 billion, there is reasonable evidence to assume that the entire American economy could save at least \$4 billion a year through standardization. Representative returns show the following examples.

A contractor specializing in petroleum and chemical works estimates savings of a million dollars a year through the use of standards, or \$8 in savings per \$1 spent on standards activities. Areas in which substantial savings were reported by the company were engineering and drafting time, purchasing, identification of materials on the job, and erection of correct parts in proper places.

A manufacturer in the missile field saves 5 percent of total sales of \$216,000,000.

A chemicals company with annual sales of \$250 million reported on its new standards program. With a total budget of \$35,000, the program resulted in the establishment of 28 standards in the first ten months, and 62 standards manuals were issued. Company personnel are active on standards committees of the Manufacturing Chemists Association, the American Society for Testing Materials, and the Society of the Plastics Industry.

A machinery manufacturer with 10 plants and annual sales of \$155 million reported tangible savings of \$250,000 a year through standards, and estimated that intangible savings per dollar spent on standardization run as high as \$35 or \$50. Its yearly standards budget is \$70,000.

Another heavy machinery maker, E. W. Bliss of Canton, Ohio, reported annual savings of about \$50,000 on a standards budget of \$14,000.

The Kentucky Utilities Company reported estimated savings of \$50,000 to \$75,000 annually as the result of a recently established standards program budgeted at \$15,000 a year. The company expects the budget to drop when the standards program is fully developed.

The St. Regis Paper Company of Providence, R.I., reported annual savings of \$107,000, and stock inventory reduction of \$1,200,000 in three years, as a result of a recently inaugurated standards program with an annual budget of \$30,000.

Savings experienced by electric and electronic companies through the use of standards range up to \$50 per dollar spent on standards activities. This fifty-fold savings was reported by Davenport Manufacturing Company of Chicago, which believes that a small company is more dependent on a solid foundation of standards than a large, well-known company that can trade on its reputation.

Western Electric Company also reports reliance on organized standards practices on a wide scale. Uniform methods are established through handbooks and manuals, and the major divisions incorporate standards procedures into their functions. In addition, many Western Electric people serve on standards committees of trade and professional associations, including the American Standards Association.

An interesting development is the indication that companies which report large savings due to their standards programs are also those that are making use of American Standards and are cooperating in national and international standardization. As the introduction to the survey points out: "ASA activity continues to increase as more and more companies, through their trade associations, through technical societies, and through other national groups, request the initiation of new American Standards projects of national scope. As companies learn of the economic benefits of standardization they learn that these benefits can be extended in many cases across company and industry lines, to the mutual benefit of all concerned."

# STANDARDS PAY DIVIDENDS

Gerald C. Joynt

In our manufacturing standards program we are concerned with standardization of production and related facilities, and with materials used in our production operations. Standardization of the product itself, and the materials therein, is a responsibility of our product engineering organization, and is therefore a separate story.

The fundamental objectives of the Ford Manufacturing Standards program are to:

Maximize employee safety
Maximize machine, tool, and equipment longevity
Promote continuity of operations
Minimize maintenance time and complexity
Minimize consumption of power, air, water, oil, coolants, and other materials used in production
Minimize stock inventories
Meet product quality requirements

I have not mentioned cost, per se, because reduction of manufacturing cost is the common factor in all these objectives. Attainment of these objectives also results in several indirect benefits: effective standardization increases the quality of production facilities and components, thus tending to eliminate inferior products. Furthermore, it promotes design simplification and interchangeability of components, resulting in greater flexibility of production operations.

### Standards Referral System

Ford manufacturing standards are approved for use through company-wide referral. As illustrated (fig. 1), proposals for company standards

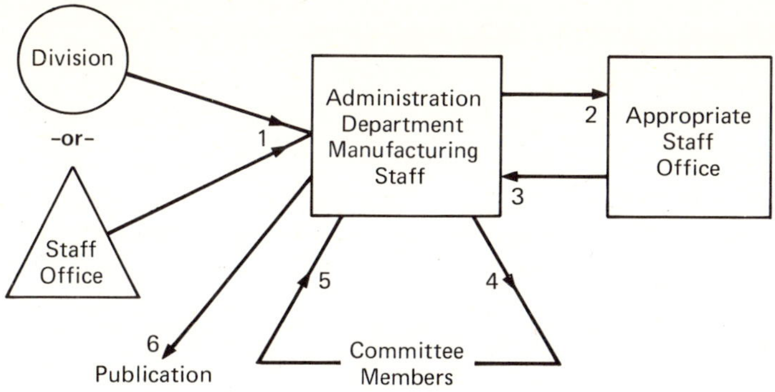


**Figure 1.** Manufacturing Standards Referral System.

originate in either a division or staff office. They are then referred, through the Administration Department of the Corporate Manufacturing Staff, to the appropriate staff office for review of technical content, and to ascertain whether the subject is suitable for inclusion as a manufacturing standard.

Approved proposals are endorsed by the staff office and forwarded to the Standards Committee, which is composed of representatives of the various operating divisions and staffs, for review and comment. Upon receipt of favorable decisions from individual committee members, the administration department codes and publishes the proposal as a company standard. In the event the proposal should be rejected, the administration department requests the appropriate staff office to secure a resolution before publication. Revised or approved new standards are distributed to all interested components of the company, and put into practice as quickly as possible.

The Administration Department of the Corporate Manufacturing Staff contains the only group in the company assigned full-time to manufacturing standards, and this is solely an administrative function. Unlike many companies, Ford has no full-time standards engineering group. We look to our operating divisions to originate proposals for new manufacturing standards. This is where the standards are needed; this is where they must be used; and this is where practice demonstrates the utility of a standard. More than 90 percent of all new or revised standards are initiated by the various manufacturing engineers in our operating divisons and plant. We expect process, material, and tool personnel to be constantly on the lookout for opportunities to further standardize the machinery, materials, and tools they work with every day. Development of new standards is one of their assigned duties.

**Example: Cutting-tool Standards**

When a staff or division engineer wishes to evaluate a particular cutting tool, he follows a definite method for testing of the item. This method is itself a Ford manufacturing standard, known as TX4. It specifies both the preparatory steps to be taken and the actual test procedure. The preparatory steps are as follows:

1. Check operations for conformity to process sheets.
2. Ascertain that machines are in good working condition, with a minimum of chatter or other poor operating conditions.
3. Install automatic counter on machine where necessary.
4. Procure tools required for test through appropriate purchasing activity. (At least three tools from each vendor should be procured for test and their identity obliterated, then coded.)
5. Log the data for each test tool on a tool performance record.

During the test, all tools, materials speeds and feeds, and other machining conditions must remain constant. Variables which cannot be avoided, or other testing methods which are necessary, are noted on the tool performance record. This also has a place to note the pieces per grind, the amount ground off the tool during regrind and minutes per regrind, the size of the wear land, chipped or broken tools, and minutes per tool change, set-up, or adjustment.

At least six test runs are made with each tool. However, to determine the feasible number of regrinds, the tools are used up. (When conducting tests we do not run a tool until it fails and will no longer cut properly; and we do not run a tool until it has become worn down so that the workpiece is no longer within acceptable tolerances. Such methods are not realistic, and will result in erroneous test results.) Tools are reground under normal conditions, using the same machine and operator wherever possible. Each tool must be accounted for at all times, and must be examined by the tool analyst before and after each regrind, with data recorded on the tool performance record.

When the test is completed, data compiled on the tool performance record is entered or computed on a tool cost analysis form, and comparative costs are determined. This data enables purchasing and manufacturing engineering to select the optimum tool, and establish company-wide standard specifications for this item.

One outgrowth of our standard tool testing program has been the

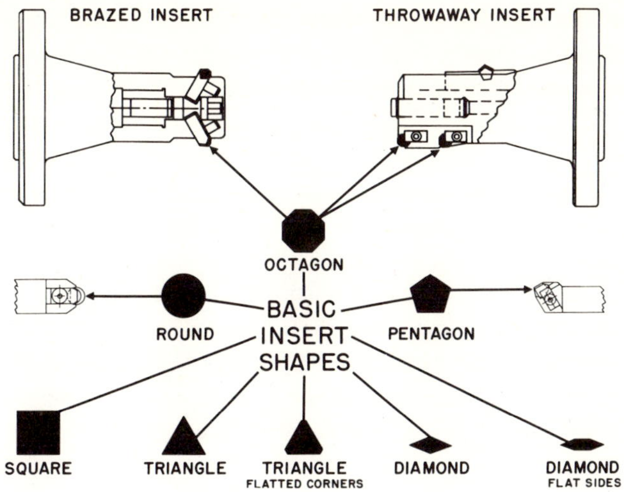


**Figure 2.** Cutting Tool Inserts.

development of cutting tool throw-away inserts. Basic insert shapes are procured in a variety of materials from suppliers who meet our standards requirements. The annual usage of tungsten carbides is estimated at 600,000, titanium carbide usage at 400,000 pieces. High-speed steel throw-away inserts are used where reduced spindle speeds are controlled by drilling operations on multispindle chucking and automatic bar machines. Some ceramic throw-aways are still used in our operation.

A specific comparison between throw-aways and brazed inserts is seen in the illustration (fig. 2). The two cutter assemblies are arranged for boring the crank end of connecting rods. To restore the cutting edge of a brazed insert tool, the tool must be removed from the machine and shipped to the cutter grind department. The blades are reset on the bench for resharpening; the assembled cutter is then resharpened to the required size, and returned to production. On the other hand, the octagon throw-away insert is provided with sixteen cutting corners. To reset the cutting edge, the insert is indexed to a new corner – on the job.

Eighty percent of all basic insert shapes are used in our Transmission and Chassis Division for turning, boring, facing, and milling operations. The round throw-away provides the maximum number of cutting edges and the octagon throw-away provides the larges number of cutting corners.

The shape of an insert required for an operation is determined by the configuration of the art and the required tool geometry. The two diamond throw-aways are used for tracer lathe operations.

### Materials

A different procedure is used to develop standards for materials used in the production process, such as coolants and plating supplies, and to qualify new materials and new suppliers. Division manufacturing engineering asks purchasing to obtain necessary technical data and samples from qualified suppliers, which manufacturing engineering forwards to the appropriate laboratory activity for analysis and checking against existing standards.

If the material appears acceptable on the basis of the laboratory analysis, manufacturing engineering conducts shop trials. When completed, a cost-per-unit comparison between the present material and the test material is established, and the accumulated data is summarized and entered on a shop trial report. This also rates the new material against the present material as equal, superior, or inferior.

Results of shop trials must be furnished with proposals for new standards. Copies of the shop trial report for accepted or rejected materials are also forwarded to purchasing and to the appropriate manufacturing engineering activity. An example of such a study is seen in figure 3.

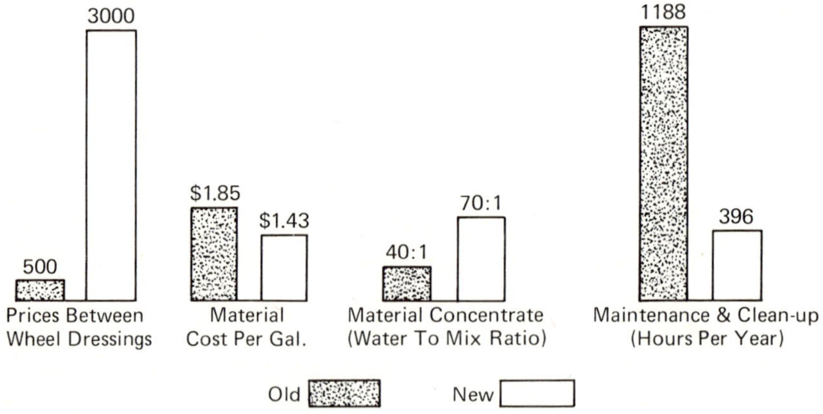


**Figure 3.** Standard Coolant Savings and Improvement.

**Standards for Machine Tools and Press Equipment**

The necessity to standardize production machine components and units was established during 1954. Dimensional variations existing in feed units manufactured by ten machine builders were charted. The lack of uniform mounting dimensions and the variation of overall dimensions were deterrents to easy vehicle model changes. In 1955 a scale model of an in-line transfer machine was constructed exemplifying the flexibility which can be attained by incorporating interchangeability in the basic design and construction of production machines. Ford interested other users and production machine builders in the project, and with their cooperation an industry standard was established. Ford's manufacturing standard was published in July, 1959.

The illustration (fig. 4) of a way-type machine depicts the basic components and units used in the construction of way-type and multiple-station production machines. It appears on page 5 of the ASA draft *Modular Machine Tool Standards*, which will perpetuate the industry standard established during 1959. The foreword of the ASA draft reads:

"This standard establishes fastening and dimensional standards to provide maximum interchangeability between Modular or Building Block components – and it will not inhibit or restrict the art of design of manufacture."

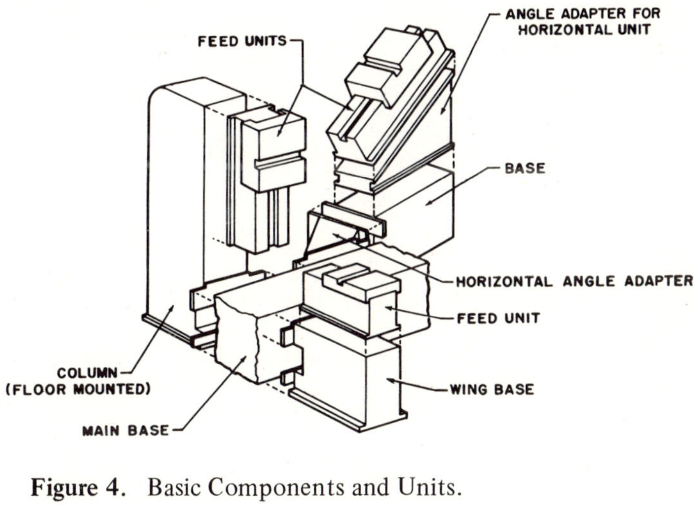


**Figure 4.** Basic Components and Units.

The manufacturing engineering office has been similarly active in the development of standardized components for stamping plant equipment, such as sliding bolsters for presses. Other areas where it has been active include new processes for cold forming, modular iron, electrical discharge machining, and finishing techniques. The research is conducted in a 100,000 square-foot facility operated by the office.

### Electrical Test

The manufacturing development office conducts a permanent program for the testing and standardization of production welding and other electrical equipment, such as guns, limit switches, and relays. Dual tests are conducted to establish standards requirements for mechanical wear and electrical life. Load banks determine electrical life, and mechanical wear life is relative, since it is determined on a comparative basis.

Limit switches and relays must withstand 10 million trouble-free cycles without adjustment or repair. The relay life test is ordinarily run from a timing box at a rate of 200 operations per minute. Limit switch tests are conducted at approximately 250 operations per minute, being actuated for 7 cycles and released for 7 cycles. This is obtained from a four-lobe cam on a shaft rotating at 62.5 rpm. The switches are also subjected to a spray of water for approximately 4 hours each day.

The welding development area covers 3,900 square feet, and employs 15 men. In a normal day, this facility makes in excess of 1.4 million welds, on many types of equipment. Although most of these are made in performance testing to establish tool standards, some welding is done on prototype and experimental parts.

Information from tests and developments results in proposals for many new and revised standards for electrodes, welding cables, transformers, welding guns, and arc welding equipment. As a result of recent manufacturing development tests –

1. Standards were revised to improve electrode and cable life, and new standards were established for electronic control panels.
2. Resistance welder transformer performance standards were established.
3. Welding gun life was increased tenfold, to over a million welds.

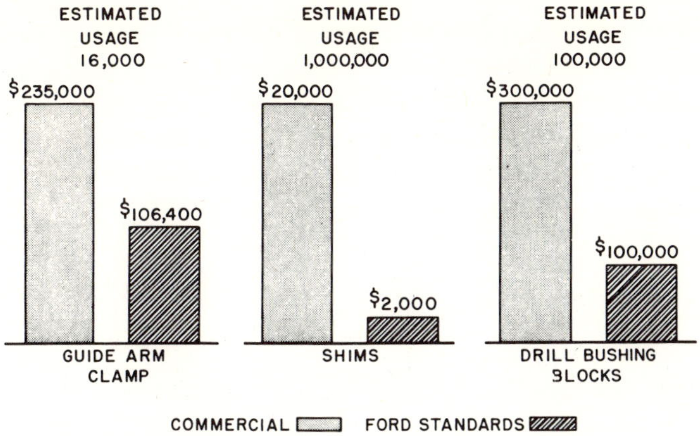


**Figure 5.** 1967 Tooling Program Cost Study Comparison: Commerical Items vs Ford Standard Items.

**Tangible vs Intangible Benefits**

Tangible benefits are defined as those benefits for which savings can be easily and economically measured. Intangible benefits are those for which savings cannot be readily measured: we know the standard is paying dividends; however, they would be difficult and costly to measure. Examples are safety and employee good will.

Some typical examples of tangible benefits are illustrated in figure 5.

In addition to saving tooling "make" costs, usage can often be improved, as seen in the next illustration (fig. 6) which shows the effect (on die setting time) of standardizing locating pins for sliding bolster plate stamping presses.

As another example of the potential in the tooling area, we determined that the estimated usage of 15 body assembly fixture components required for the 1967 model tooling program was more than 1.2 million pieces. The total cost of these pieces was estimated at more than $2,300,000 before standardization. As a result of standardizing, the part cost dropped to a little over $1,100,000. See figure 7.

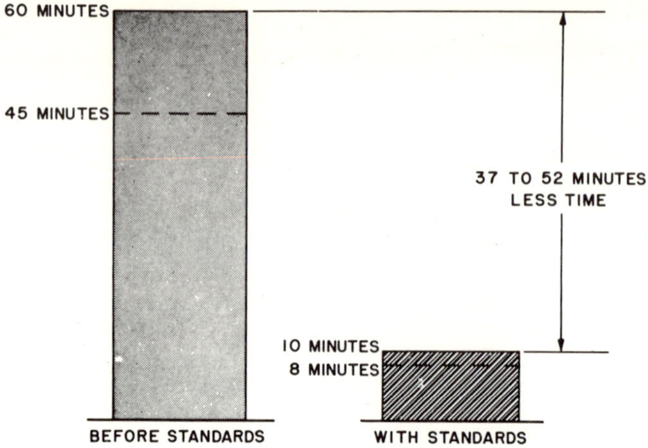


**Figure 6.** Locating-pin Standardization.

The following are examples of standards that result in intangible benefits.

1. There has been a tremendous reduction in accidents due to standards on electrical, hydraulic, and pneumatic controls on production equipment. As seen in figure 8, the frequency of accidents per million man hours was reduced 92% from 1947 to 1964. (The accident defined in this category is an injury that results in one or more days' lost work, or a permanent injury.)

2. A hand guard for an electric truck was standardized in 1958. After 3 years, it was found that accidents to the operators' hands were down 75 percent.

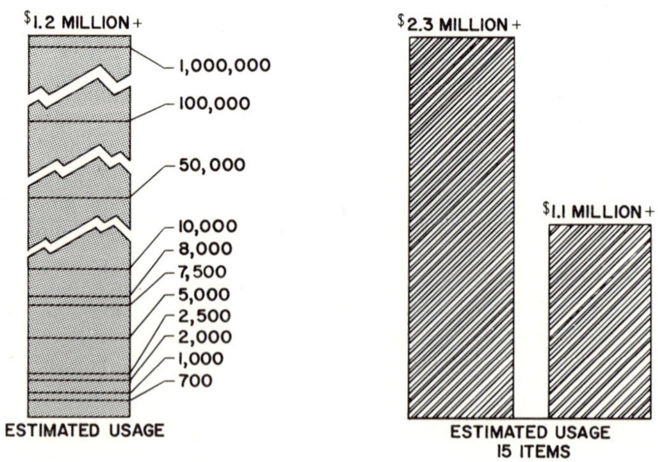


**Figure 7.** 1967 Model Tooling Program.

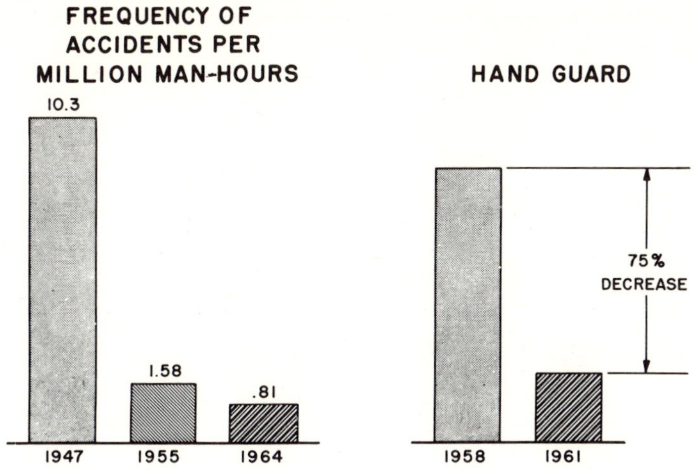


**Figure 8.** Savings in Safety

Historical records thus indicate fewer accidents as a result of standardization. We knew that savings result from lower insurance rates. These savings are difficult to measure, but we can evaluate them in terms of more, and less. The more effective the standards program, the fewer the accidents.

In another area, we remember that, prior to standardization, industry was fortunate if it could get 132,000 cycles from control devices such as relays and limit switches. Today, controls that have been improved through our testing and standards program now have a life expectancy of at least 10 million cycles. While it is difficult to measure the savings which have resulted, it is apparent that downtime and maintenance are greatly reduced. The machine or equipment is in service performing its intended function, and is making money rather than costing money for replacement items and maintenance time.

Our Buffalo, Cleveland, and Chicago stamping plants have standard electrical control panels for presses. As a result, presses may be interchanged between plants; the men are not required to learn new circuits; and replacement parts are interchangeable.

Fastening patterns for holding dies in OBI presses may be standardized. Any die will fit in any OBI press in any stamping plant within the division. The same type of interchangeability applies for intermediate size presses, and major presses. Here again the savings would be difficult to measure; however, interchangeability and flexibility are achieved, simplicity of operation is improved, and changeover and set-up times are reduced.

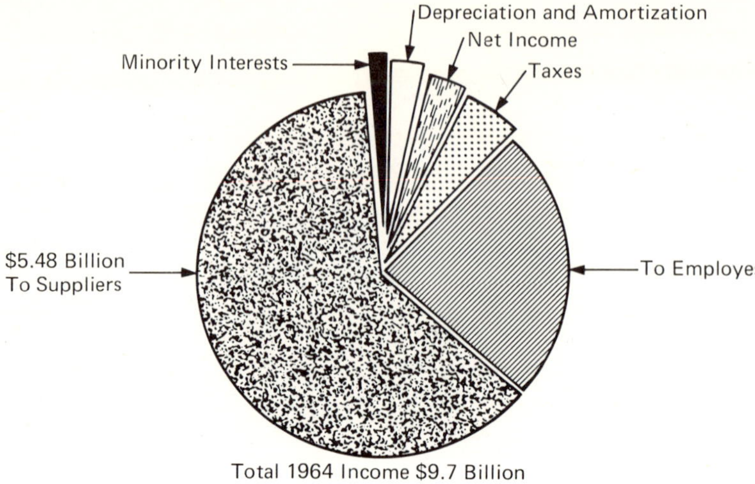


**Figure 9.** The $5.48 billion going to suppliers provides fruitful opportunity for cost reduction through standardization.

**In Summary**

According to Fords' 1964 Annual Report, the distribution of sales and other income totaled almost 9.7 billion dollars. The biggest cut of the pie – almost 5.5 billion dollars – goes to suppliers. See figure 9. This area provides opportunity for cost reduction through standardization. Each new model program compounds the potential for standardization. At the same time, it is always possible to further standardize existing operations.

In each instance, the need to standardize should be objective and initial selection of items for standardization should be based on sound judgment. The steps should be:

1. Standardize on commercially marketed items wherever possible.
2. Standardize on other noncontroversial items.
3. Establish programs to review controversial items.

Most of the items proposed for standardization are noncontroversial in nature, such as commercially marketed items. These are identified and expedited through the referral system. Inevitably, however, certain proposed items are bound to be controversial. Where it is anticipated that significant differences of opinion may be encountered regarding design or other characteristics, we attempt to reach basic agreement among the interested activities before the standard is officially proposed through the referral system.

Our objective is to maintain standards that are compatible with the latest technical developments. This requires the closest possible liaison between process development engineers, manufacturing engineers, and suppliers. All remember that an obsolete standard is worse than no standard at all.

Reiterating the basic principles which we constantly strive to apply in our Manufacturing Standards Program:

1. We take a "horse-sense" approach, applying this question to each proposed standard – "Will it increase the efficiency of operations under actual production conditions?" We know that it is possible to overstandardize. Standards are only a means to an end, not an end in themselves.

2. We try to make sure that a standard will not restrict the art of design, or limit further improvements and use of alternate materials. We want nothing to inhibit design creativity, technical innovation, or new ideas.

3. Our manufacturing standards must implement, not impede, our progress toward greater flexibility of production facilities. To meet the demands of today's market, we have to be able to run many different components and models through the same manufacturing and assembly facilities.

4. Adequate testing is a "must" if standards are to meet expectations. Comprehensive test programs on tools and materials inevitably involve supplier firms which often invest considerable capital in a cooperative program. Such aid must be fairly recognized and fairly compensated.

5. Like anything else, a standards program must have the firm backing of higher management, and adequate recognition of individual accomplishment, if it is to endure.

Ford's Manufacturing Standards Program has endured, thanks to the continual efforts of the engineering people in our plants, divisions, and staffs. Their interest and enthusiasm have paid off in higher quality and lower costs.

# THE TANGIBLE BENEFITS OF STANDARDIZATION

W. F. Arnold

How many times have you heard, "The benefits derived from standardization are intangible?" Standards engineers should take exception to this word "intangible" – the benefits derived from standardization not only are tangible, but also, in many cases, they can easily be measured.

Reduction of cost, consistency of design, and reduction of development time constitute some of the more obvious tangible benefits from standardization. See figure 1.

## Reduction of Cost

The average cost of releasing a new part has been established in standards literature at between $1,000 and $2,500, depending upon the size of a company's engineering, manufacturing, and field organizations. Although these figures are often questioned they can be proved by considering these expenses (see also fig. 2):

1. Engineering and drafting time required to prepare original documents
2. Cost of maintaining engineering and manufacturing records
3. Required storage space for records
4. Tool analysis and tool costs
5. Production control

Promoting the reuse of existing design and parts standards reduces costs by: (1) reducing the number of new parts released; (2) reducing the

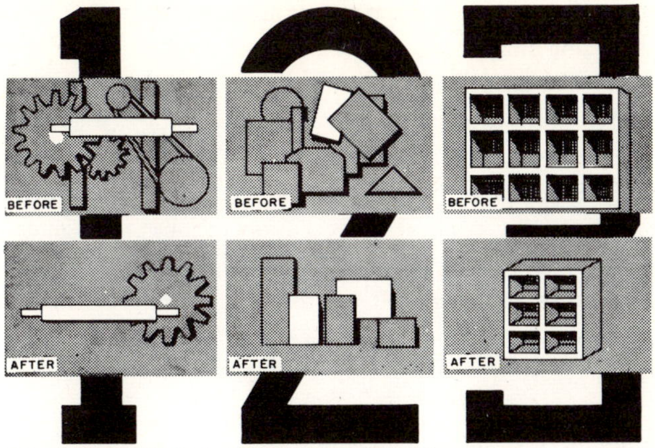


**Figure 1.**

size and variety of materials required; and (3) reducing parts inventory. Reuse of the same part also increases the production volume of the part, and thereby directly lowers its cost.

## Consistency of Design

Standardization helps provide design consistency and encourages reuse of proved components by aiding in their selection as well as by

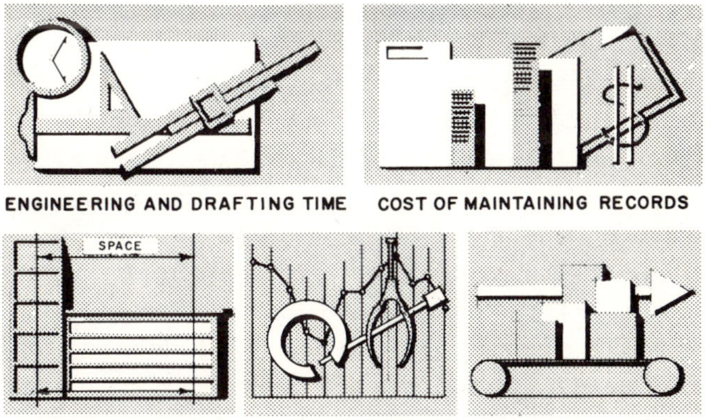


**Figure 2.**

establishing practical limits on design techniques. Selection of materials, methods of assigning dimensions, tolerances, limits and fits, and other design standards significantly affect the manufacturing cost of a product. Imagine the variety of tools, inspection gauges, and fixtures that would be required if each engineer were permitted to follow different procedures and design limits for similar applications. Utilizing design standards results in reuse of existing tools, reduces production operations and engineering changes, minimizes potential scrap and rework charges, and simplifies inspection procedures. At the same time, the design is improved because of better quality control on purchased and manufactured parts.

### Reduction of Development Time

Standardization can also directly reduce the time required to go from development to production. It has been estimated that as much as 30 percent of the effort in a new product development program is a duplication of what has already been accomplished in previous programs. This means that it is not feasible to evaluate a production model with a view to standardize "after the fact."

### Implementation of the Standards Program

How is the standards program effectively implemented? Every standards organization should be responsible for determining which components will be recognized as "standard." Selection of these components should be based on reliability, economy, and interchangeability. After the components have been selected, they should be assigned "family codes" for accessibility. Lists of these components should be published and made available to all engineering areas.

The standards organization should also maintain and publish a list of satisfactory nonstandard items. Use of nonstandard, but previously available, approved items, should be encouraged where a standard item does not suffice, since this is more economical than forcing the release of a new item by restricting information on satisfactory nonstandard parts.

Nonstandard item listings, however, are of little value if the parts are not actually used in design. Standardization presents a conflict of interests to the engineer because by virtue of his training the engineer is encouraged to be creative and original. His success is, to some degree, measured by the number of new and different things he designs. This, of course, introduces the "NIH" factor – "not invented here." Standardization, however, should never suppress the creativity of engineering; it should instead stimulate the tendency to originate economical designs. This places an additional responsibility on the standards engineer – motivating the designer to use the standards data available.

The first step in motivation is to select standard parts that satisfy the needs of the company; the second is to make designers account to management for use of nonstandard parts where standard items are available. The designer's performance should be based not only on his contribution to the state-of-the-art and in production of new designs, but also on the basis of total design cost and his contribution towards net profit. To measure progress, the cost of introducing new items and techniques must be subtracted from the benefits derived and weighed against the possible advantages of using standard parts and techniques.

The standards engineer has still another responsibility. He must be capable of selling his ideas and his products. He must sell to the new engineer the idea that existing design information is an asset in development programs. He must also sell management on the idea of extended standards activity, emphasizing standardization "before the fact." Once management accepts this philosophy, it is easier to convince the engineers. The end results are economical, reliable, high-quality products.

What assurance does a company have that engineering will utilize design standards? None, unless a standards group assists in the use of these data. The application activity has been labeled as "policing" by those who think of standards as a "necessary evil." This attitude must be overcome if standardization is to benefit industry.

At IBM, engineering standards generally are recommended practices and are not mandatory except in the case of standards for operating noise level, machine size, product safety, and power requirements. Many IBM standards are extracted from American Standards Association (ASA) publications and rewritten to fulfill our requirements. Before an article can be published as a standard, each plant within the division must approve the contents.

Standards engineers must be in direct contact with engineering groups to obtain a more comprehensive understanding of engineering problems. This aids in upgrading existing standards and developing new standards. Every machine program should have standards representatives assisting in the preliminary development stages and just prior to production release. This procedure minimizes deviations from standards and expedites release to production; and the standards engineer acquires a broad perspective for the application of standards and gains the confidence of the engineers, thereby eliminating the "necessary evil" obstacle.

Cost data are established for every engineering and manufacturing operation, which makes it possible to assign a value to design standards. Consider the hypothetical example of a part failure:

1. The engineer insists that he must deviate from a design standard because it is too restrictive or too lenient for his application.
2. New manufacturing tools or fixtures, or both, must be acquired to produce the product.
3. The components are assembled in the machine, and a failure results.
4. The problem is traced to the nonstandard items and it is determined that compliance with the design standard will correct the condition.

The cost of each operation – to obtain the original parts, assemble them in the machine, and trace the problem – is available. To correct the problem, there must be an engineering change, possible scrap or rework charges, and much duplication of effort. All of these operations have dollar values to the company and must be considered tangible.

A few years ago, an attempt was made to affix a value to IBM's parts listings. To accomplish this, engineers and designers were asked two questions:

1. "If standard and common parts listings were not available, what percentage of existing parts used in previous designs would you remember and apply in future designs?"
2. "What percentage of parts do you find in the standard parts listings that are satisfactory for your new design requirements?"

Answers to the first question indicated that approximately 15 percent of the parts would be remembered. The actual reuse, however, would depend on the function of the design. Answers to the second question indicated that approximately 40 percent of the parts selected from the listings were applicable in a particular design.

Fortified with the results of this survey, a parts analysis was initiated for an actual machine. The results are shown in table 1.

**Table 1. Parts Analysis after Standards Review**

| *Standard Part Numbers* | | *Noncoded Part Numbers* | |
|---|---|---|---|
| Existing | 397 | Existing | 380 |
| New releases | 5 | New releases | 785 |
| Total | 402 | Total | 1,165 |
| *Nonstandard Coded Part Numbers* | | *Total Part Numbers Required* | |
| Existing | 751 | Existing | 1,528 |
| New releases | 515 | New releases | 1,305 |
| Total | 1,266 | Total | 2,833 |

The engineer's estimate of 15 percent recall for existing part numbers seems conservative when analyzing the application of "Noncoded Part Numbers" in table 1. Of the 1,165 noncoded part numbers used, 380 were existing numbers recalled from memory, or 33 percent. The design of noncoded parts is so varied that the parts cannot be categorized by family codes, consequently their reuse should be limited in comparison to the coded parts. Therefore, it is only logical that a greater percentage of the more commonly used parts (coded) would be remembered if listings were not available.

Assume a hypothetical case of no standards activity whatsoever, i.e., no parts listings, technical guidance, etc. All reusable parts on a machine would be recalled from memory.

Use the statistics of table 1 as the hypothetical case and apply the following memory factors: 33 percent for noncoded part numbers, 50 percent for coded standard and nonstandard part numbers, thus:

| | |
|---|---|
| Existing standard part numbers | 402 X 0.50 = 201 |
| Existing nonstandard part numbers | 1266 X 0.50 = 633 |
| Noncoded part numbers | 380 |
| Total part numbers recalled | 1,214 |

Out of 2,833 part numbers required, 1,619 would have been new releases. In comparison, the results of table 1 show that 1,305 new releases were required after standards' review.

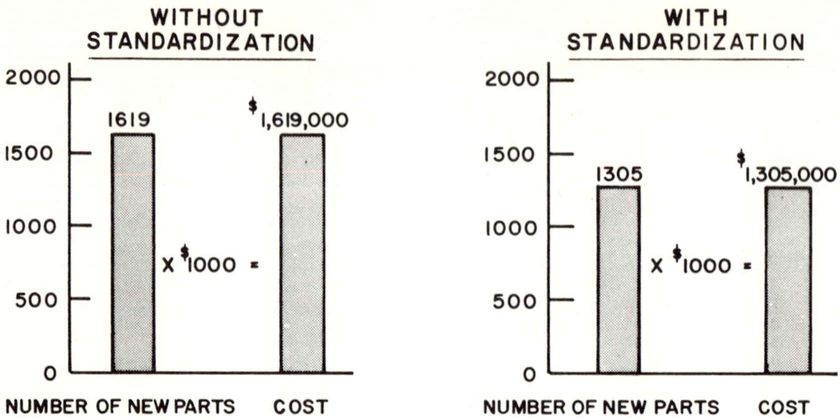


**Figure 3.**

Since the cost of maintaining parts listings has not been considered, the minimum cost of releasing a part ($1,000) will be used to assign the dollar value. The cost of releasing new part numbers in the hypothetical case is therefore $1,619,000 (1,619 new numbers x $1,000). The cost of releasing new numbers after standards' review is $1,305,000 (1,305 new numbers x $1,000). This means that the standards activity represents a parts savings of $314,000. See figure 3.

Although a hypothetical case has been assumed for the comparison, this example proves quite conclusively that standardization not only offers tangible benefits to industry, but offers benefits that can be measured.

# ADVANTAGES OF STANDARDIZATION TO PROCUREMENT ACTIVITIES

V. deP. Goubeau

Having come to appreciate the importance of standardization based on experience in procurement over a great many years, I deem it a distinct privilege to have the opportunity to discuss this subject with you from a purchasing man's viewpoint.

I am impressed with the background of the Standards Engineers Society as described to me by Mr. Gokhale. It indicates that you gentlemen are not only interested in furthering the work of standardization, but are ready to give of your personal time and energy to insure the program's success. I often think of one of the pet expressions of our president, Mr. Frank Felson. He states that the people who do worthwhile things in this world usually do them because they want to and not because they have to. You are making progress in your society through the application of this principle.

The progressive purchasing agent of today has the desire to improve his effectiveness by exploring new fields, new approaches, and new areas, which will introduce more scientific procurement and fewer hit-or-miss day-by-day buying routines. Purchasing men are more and more becoming aware not so much of the importance of standardization itself – they have been aware of that for a long time – but of the importance of actively furthering its use. This objective might well be called a new horizon of purchasing.

Every day we are doing more and more about the adoption of standards, but there is still much to be done, and it seems to me everyone involved, whether in specification writing, engineering, procurement, or handling of materials, has a part to play in hastening further progress.

Just as a little background, I would like to review the activities of the National Association of Purchasing Agents in connection with this subject. You will recall that about three or four years ago, Tom Jolly, who is vice president in charge of engineering and purchasing for the Aluminum Company of America, was president of the American Standards Association. He and I discussed the desirability of having the National Association of Purchasing Agents take some active part in the work of ASA. Tom Jolly sold the idea to the Executive Committee of the National Association of Purchasing Agents and, as a result, NAPA became a member body of ASA, and a committee on standardization was formed. It was agreed at the outset that we would not attempt to promulgate standards – that is a job for the technical societies – but we would, as a group, assist and aggressively pursue the objective of progress in standardization.

During these past three years, we have had several forums on the subject in NAPA meetings as well as in ASA sessions. At the ASA meeting in Chicago last September, representatives from the Company-Member Conference of ASA and members of NAPA met together, and we are planning a purchasing forum at ASA's annual meeting here in New York in October.

The recent Annual Convention of the National Association of Purchasing Agents, which took place in Los Angeles May 25 to 27, 1953, included a forum on standardization, and the excellent attendance was very encouraging. The nature of the questions asked by the members demonstrated clearly their interest in becoming a part of this increasingly important phase of industrial management. Many illustrations were given of dollar-and-cents savings in the cost of materials through the use of standards rather than specials.

The effectiveness of any committee is only realized when the program is picked up and activated at the local level. Therefore, it has been heartening to our committee to note recently that the local chapters which go to make up the National Association of Purchasing Agents are introducing standardization programs in their activities, which should go far to make for real progress.

Purchasing men over the years have been suffering from the handicaps of buying specials. Every experienced purchasing man knows the headaches of using special materials rather than standards. Your inventory builds up if you use a different item for every purpose, rather than one item for many purposes. Every purchasing agent knows that the cost of a special item is higher because the volume is restricted. In addition, the number of suppliers from whom he can obtain a special is limited, thus reducing competition. The purchasing agent recognizes the ever-present danger of obsolescence that lies in the use of special materials. Whether he realizes it or not, therefore, he probably has been playing quite a part in promoting the use of standard materials rather than specials. The only remaining step is to direct his activities more systematically – to put them on a programmed basis instead of an informal, hit-or-miss basis.

Despite the constant interest of purchasing managers in the use of standards and the avoidance of specials, some interesting revelations have come to light since NAPA formed its standardization committee. Some members were at a loss to know specifically what they should do about the subject. This can be easily understood, considering that it is such a broad one and covers such a variety of fields. Therefore it was only natural that purchasing men would ask, "What should I do about standardization in my own company?" I have been asked frequently, "Should I as a purchasing agent establish a standardization operation?" Usually our answer is that if you already have provision for standardizing in your company, such as might be carried on by the engineering department or a standardization division in an engineering department, as we have in RCA, it is not wise to establish a standardization function in purchasing because the use of materials is a matter of specification and usually specifications are prepared by the engineering department. If engineering is responsible for the design of a product, it should be responsible for the specifications which go to make it up. The purchasing department, however, can be of tremendous assistance in this field by bringing to the engineering department current information on costs, availability, and performance of standard items which might do the work of the specials which have been specified.

We have further suggested that the most satisfactory approach to proper collaboration usually is to develop a standardization committee, composed of both engineering and purchasing representatives, as well as any others who might be interested. If the company has a standardizing division, it is well for this division to spearhead the work of this committee, with purchasing representation playing a supporting role.

While it is most important to pursue aggressively the use of standards rather than specials, I thoroughly realize that proper balance at all times is necessary, and that we must not stand in the way of progress. For that reason, I believe it most important that we maintain an open-minded attitude. If, in our efforts to avoid the use of specials, we create a reactionary atmosphere – if we are so restrictive that we in turn restrict the vision of our engineers – we have defeated our purpose. In such an atmosphere, radar, guided missiles, and other advanced products of today would have been suffocated in their cradles.

The word "standardization" is a mighty broad term, covering many different activities from standardization of drafting practices to purchasing methods, and so on. In keeping with our subject tonight, I would like to discuss primarily the purchasing executive's interest in standards as related to the things a purchasing department buys. This might be broken down into two principal areas: materials and components used in assembling the product, which we call production items; and items in the nonproductive area such as shop and office equipment, maintenance materials, etc.

The purchasing department has the responsibility of obtaining the right materials of the right quality, in the right quantity, at the right time, and at the right price. These are the four cardinal principles of purchasing. All four principles involve the joint responsibility and work of others. Quality is truly related to specifications, so that the finished product will perform as planned. The quantity has to tie in with sales and factory plans, inventory objectives, and the decision of management as to how much money to tie up in inventory. The time of delivery is related to the needs of the factory, to permit shipment of the product on schedule. Price especially is of interest to other functions of the company because the cost of the product will have to coincide with the ability of sales to sell at a profit. Now, while it is the duty of the purchasing department to fill requisitions in accordance with these four principles, it also has the further broad responsibility of continually seeking items which will represent improvements – whether in lower cost, or improved quality, or better efficiency or what-not – and thus contribute to the ultimate goal of the company, which is to make a profit.

It often happens that a special item is used through habit or because of an earlier original design. This is the engineer's final responsibility, of course, but it is the duty of the purchasing man to bring to the attention of the engineer information showing how savings might be obtained through the use of a standard item. The engineer may not realize, for example, how great is the difference in cost between the item specified and the standard one.

Purchasing men, by virtue of their constant contact with the outside world, very often are in a position to point out to engineering or to the standards engineer materials which may make for excellent substitutes at lowered costs. On occasion I hear complaints from suppliers who claim that they present their wares to buyers only to have them die in the buyers' desks. That type of buyer cannot live long. It is the responsibility of the buyer to be ever-attentive to the availability of new materials, so that he may present these new items to the engineering people for their consideration.

The General Electric Company has done quite a job with a subject called "value analysis." It is nothing more than a study of the items being used to see if, through improved processes or substitute materials, shortcuts in manufacturing, or other approaches, lower costs may be achieved. A plastic plate or washer may make perfectly adequate substitute for brass. And while this example may not come under the narrow heading of standards, it illustrates the buyer's contribution to the lowered cost of the product without sacrificing quality.

If the buyer proposes such a substitute to the engineer, and it proves to be acceptable, their joint effort may very well result in its adoption as a company standard.

We have found on occasion that we were buying several different qualities of a component for use in assembling our various products. The cost of purchasing limited quantities of these several different grades resulted in considerable handling and higher cost of stockkeeping. In many cases, analysis showed it would pay us to standardize on the best quality of an item for all uses. The net overall cost to the company was the same or lower because of the resulting simplicity in handling, purchasing, stockkeeping, and issuance. For example, we found this to be true of resistors used in our engineering products department where we manufacture commercial electronic items rather than consumer products.

With regard to the purchase of nonproductive items including equipment, materials for maintenance needs, office furniture, and the like, we have made very satisfactory progress in the adoption and use of standards. There were many problems involved, one of the primary ones being the individuality of the maintenance or plant engineer. One engineer considers a certain brand of lubricating oil to be the "best in the world," while another swears by a different brand. In multiple-plant operations, you will find similar tendencies toward personal preferences in materials for maintenance, whether paint, oil, material handling equipment, or what-not. If you are buying paint for fifteen plants and you find yourself purchasing from fifteen different companies because of individual tastes, you must conclude that all fifteen can't be right – there must be one, two, or a few best products to use across the board.

It is not uncommon to find a great deal of attention being given to the materials going into a manufactured product while the economical purchase of maintenance items is neglected; yet there are substantial dividends to be derived from an intelligent standardization program in this area. In our own company, through a class team arrangement between plant engineering and purchasing divisions, great savings have been made through the adoption of standards in the maintenance field. When objections were raised, most of them were found to rest on personal pride or prejudice and consequently could be eliminated by frank discussion.

I had occasion recently to visit one of our plants and heard a long, sad tale about the unsatisfactory office furniture on which we had standardized. When I explained the advantages accruing to the company from the adoption of these standards, the experience of our many plants, the simplicity with which we now were able to handle the purchase of these items, and the money we were saving doing it, a few faces turned rather red and we had no difficulty in reaching an acceptance of our program. I must admit that perhaps in this case the standards had been adopted without satisfactory explanations so as to arrive at a united viewpoint, and that the purchasing division may have given the impression of saying "this is what you will use from here on out."

It seems to me that the team concept is a very important one in connection with progress in standardization. I realize that human nature being what it is, obstinacy will be present always in some degree, but it can be minimized by the process of free expression. Most of us will go along with a project if we are brought into the discussion – we are least inclined to agree when something is foisted upon us without our having been given an opportunity to play a part in its discussion or adoption.

As we all know, the use of standards at the earliest possible moment in the design of a new product saves tremendous headaches later on. For that reason we urge purchasing men to work with the engineers whenever possible in the earliest stages. The product on the drafting board today may be the item of volume production tomorrow. If a model has been made up of special materials and components, it is mighty hard to take them out of the design when the time comes for full production. It seems to me that you standardizing engineers should leave no stone unturned to see that the design engineer avoids the introduction of specials in any new products unless he is completely convinced that a standard item cannot do the job. The design engineer should be just as interested in the ultimate cost of the product as anyone else. Consequently, when a possible saving is properly brought to his attention, he will normally cooperate to the fullest extent.

Another problem we have to face continually is the desire to change specifications. After the completion of the first lot and preparations are being made to produce a second lot, it frequently happens that somebody decides to improve the product. If we are not careful, specials may be injected at this point and the first thing we know our costs of production have risen, perhaps to the point where the article is no longer profitable or competitive.

You are engaged in a very vital phase of the industrial process. Were it not for standards, we would be far down the ladder in progress, and never would have built up the great industrial machine which played such a vital role in winning the Second World War.

As I suggested before, and at the risk of some redundancy, the purchasing agent, whether he realizes it or not, has contributed substantially to standardization by avoiding specials in favor of standard items. His contribution necessarily is fragmentary and consists of hits on targets of opportunity. When buying starts, it is too late to double back and introduce many of the standard items which might easily have been incorporated at an earlier stage. The purchasing agent, therefore, looks to you men in the standards engineering field for the comprehensive, scientifically planned standards that will keep specials to an absolute minimum and at the same time keep the door open for the use of specials where they are justified – in fact, where they are absolutely essential if the progress peculiar to American industry is to continue.

# STANDARDS EVALUATION FOR PROFIT

R. L. M. Rice

I assume that we are all in agreement that our standards work is worthwhile only if it pays off. There is not as much agreement on the way a program should be evaluated to measure this payoff. Perhaps this is because the payoff can take many forms, from a direct dollar increase in your companies' profits to such intangibles as better relations with customers or suppliers.

**New Standards Programs**

From the very inception of a new standards program, we must be concerned with evaluation. But embarking on an engineering cost reduction venture of this type requires that we make an economic study before a new program is established.

Two factors must be understood: first, you cannot anticipate an immediate return on your investment; and second, you must be prepared to make sustaining expenditures to maintain the program.

*Time Element*

Proper selection and preparation of material – and obtaining reasonably broad coverage on any particular subject – takes time. At Du Pont, we find that it is not unusual for two or three years to elapse before standards work in a new field receives general acceptance. In some cases it has taken this long to develop a single standard, although this is the exception.

*Sustaining Expenditure*

It is a common error to assume that a program can be established for a limited time and that the standards produced will then be used indefinitely. With the present rate of technological development, both in materials and in methods, a sound maintenance program is necessary. Otherwise, the standards soon become obsolete, and as soon as a user finds one obsolete standard, he may be suspicious of other standards and be reluctant to use any of them.

The justification of an adequate annual budget for maintaining standards and developing new ones is rightfully the responsibility of standards management, and this justification is actually the end result of sound evaluation procedures.

## Du Pont's Program

The best way for me to introduce standards evaluation is to describe our standards program at Du Pont, with a few examples of the way standards are saving us money.

Let me give you a little of the background. Du Pont standards activity was under way in the early 1900s. We have a copy of our standards issued in 1911. They have grown from that small volume containing specifications for a few common equipment items used by the central engineering department, to about 3,500 standards relating to plant buildings, equipment, and supplies. These are used throughout the company for the design, construction, and maintenance of Du Pont facilities and for procurement of materials and equipment.

Du Pont standards were sponsored by the engineering department until 1947. By that time they had become so widely accepted in our 80 or so plants that it was deemed advisable to have policy decisions made by wider company representation. Today, the Engineering Standards Committee is composed of representatives from all 12 of our manufacturing departments, and has 8 members from auxiliary departments and one subsidiary company. This committee is responsible for the guidance of 30 subcommittees developing standards in 21 fields of engineering. There are 250 members on these subcommittees, all competent in their particular fields of interest. All but a few of these subcommittees meet once a month to conduct their business.

The selection of the subject material for a standard and its initial development is based almost entirely on the opinion of the subcommittee members involved. The cost of making a formal financial analysis at this state usually is out of proportion to the cost of the development work. Not until the information is published in a tentative form do we have an idea of its general acceptance. Approximately 100 copies are forwarded to people all over the company who may be interested in the particular subject matter. Their comments, plus their estimate of uses and savings, have a direct bearing on final publication of the standard.

The objective is to reduce to a routine those engineering tasks which are repetitive in nature. What is routine? The design of stairways in 20 mill buildings is routine. The consultant's answer to 10 identical inquiries from that many plants is routine. The methods used in supporting $20,000,000 worth of piping each year are routine. The maintenance of recorder-controllers on 80 plants is routine.

In your company, perhaps the selection of bearings or fasteners for a new machine is routine. Or, the choice of components for an electronic circuit, or the methods used to paint or finish a product are routine.

Standards should serve to improve each engineer's effectiveness. They contain an accumulation of company know-how and experience regarding correct, economical ways of doing repetitive jobs. Anyone taking advantage of this know-how can do these more-or-less routine, repetitive jobs more quickly and easily and devote most of his energy, knowledge, and talent to the special problems inherent in those jobs which cannot be covered by standards. The use of our standards is not mandatory; however, we do expect that they will receive first consideration in any given situation. We do not intend that the standards should in any way restrict the individual initiative of our engineers when this initiative is required to solve special or nonstandard problems.

### Examples of Savings

1. In the design and procurement of tanks, 8 to 10 standards and specifications are used at one time or another. Each order usually is accompanied by 4 to 6 standards and at least one specification. These standards cover details such as the design of pedestals, leg supports, flanges, nozzles, and manholes. Specifications cover fabrication details, materials of construction, and methods of shipment. Tank drawings are prepared making reference

to these standards, thus saving engineering time which would be spent developing and detailing this information. The standards then accompany the inquiry, permitting the tank vendor to include the details on his shop drawing.

It has been estimated that the standards save at least 2 to 3 man-days in the design and ordering of each tank. On the purchase of 700 to 800 tanks per year, the savings amount to a sizeable sum.

2. We have a series of standards which permits the selection and design of pipe hangers for any project to be done in 1/10 the time which would be required without standards. Since piping represents a major portion of the cost of any new project in the chemical industry, these standards contribute a large cost reduction.

3. Many of our specifications not only save the engineer time, but also result in better quotations by providing adequate information to the vendor. A specification and procurement worksheet for bellows-type expansion joints reduces the designer's time by up to 8 hours as well as ensuring that the vendor receives all necessary data. Blank copies of these forms are available from the standards section. Other types of equipment, such as pumps, heat exchangers, and distillation columns, are covered in a similar manner.

4. In the building area, architectural and electrical standards provide outstanding opportunities for cost reductions. One standard covers a sliding door installation. Details are shown for track, sill, jam, and lintel. Each use of this standard can save up to 8 hours drafting time by eliminating the need for details on the blueprint.

5. Motors in industrial locations often are subject to moisture, dust, and chemical fumes in their operation. An electrical standard aids in selecting a motor enclosure to suit the environment and covers in detail the minimum type of motor enclosures suitable for both nonhazardous and hazardous locations. Included is a graph showing approximate cost of 4 types of motor enclosures. In the past, many of our plants have felt that the totally enclosed fan-cooled motor should be purchased as standard to assure reliability and prevent application of inadequate enclosures in marginal areas. The standard points out a number of insulating treatments available today which can produce longer motor life, and also mentions several types of open-frame motors available now for installations that would have required a totally enclosed fan-cooled motor in the past.

6. Another electrical standard covering aluminum conduit contains a cost analysis evaluating different types of conduit. Curves show comparative material cost and installation cost for several kinds of conduit.

7. Most of our plants participate actively in the standards program and have developed important cost reduction items based on the use of engineering standards. An outstanding example at one plant resulted from a study of valve usage and purchases. The plant concluded that reconditioning old valves could result in large savings over the purchase of new ones.

Reconditioning valves costs the plant less than half the new valve cost; and by upgrading the packing and seating materials, when the valves are rebuilt, to conform to the latest standards, the plant actually is getting longer life (3 or 4 to 1) from the reconditioned valves. By using engineering standards as the basis for their valve reconditioning program, the plant has achieved savings of over $100,000 per year and a 50 percent reduction in spare parts inventory.

### Report Your Savings

These are some of the ways we save money through standards. There are many different ways in which these savings can be described and reported. However, it seems to me savings reports are a "must," to make sure we are getting the most out of standards and to insure continued management backing for the program.

We don't have to make much of an analysis to realize that many standards programs are not being evaluated as well as they should be. In 1955, and again in 1959, the ASA published the results of surveys of company standards programs; and in both cases the conclusions drawn were about the same. The introduction to the 1959 report discusses the answers from various companies to questions regarding the extent, scope, savings, and cost of their standards program:

> The most conclusive finding was that American industry as a whole probably doesn't know the answers to such questions. About 2,800 companies were queried, and 238 questionnaires returned, of which 209 reported varying degrees of standards activities. Of these, 89 reported a formal, organized program. Thirty-four of the 89 were able to report actual dollar savings . . . The crucial finding of the survey is that 30 of the 34 companies reporting dollar savings also report extensive use of standards and standards manuals. This seems to indicate that there is a close link between organized standards work and recordable cost savings.

Du Pont's approach to this problem is not especially complex. We have always had to justify standards on a dollars-and-cents basis – in the early days when we spent a few thousand dollars on the program, and today when our operating budget is in the neighborhood of one-half million dollars a year – and reported savings by the users of the standards amounts to several millions of dollars each year.

To evaluate our program, we send, once a year, a questionnaire to about 300 key supervisory people throughout the company. These include works engineers, laboratory directors, design project managers, and, in construction, various field project managers. The questionnaire asks these supervisors how often their people use the standards and what savings they estimate result through their use. When these forms are returned to the standards section, we merely add the dollars shown to obtain a gross savings figure which is then reported to management.

The results of a typical survey may be of interest. In 1961, the questionnaires indicated that standards were used by almost 6,000 technical people about a million times. After correcting for federal taxes and all other expenses we estimate a net return of more than $4 for every dollar invested in the program.

### Cost Reduction Programs

The evaluation of standards can be integrated effectively with other company cost reduction programs. Our engineering department encourages creative thinking, not only in research and development, but in more economical designs of equipment and methods of construction. At every site on which we function, a suggestion plan is in effect whereby an employee is rewarded for a cost reduction idea which is beyond normal job expectancy. Standards can be used as criteria for measuring suggested improvements.

The various divisions of our engineering department are competitive in reporting cost reduction activities. These are on-the-job cost-saving ideas measured against previous or normal practice. Credit is acknowledged in this program for updating and developing new standards. Credit is also acknowledged in the use of these standards in making economies. Savings reported through this medium are considered to be basically material and labor, in contrast to savings reported through the annual survey which are believed to be essentially engineering time.

### Intangible Savings

In certain areas of our operation, it is not always possible to be so factual – to present dollars-and-cents savings. Every standards engineer has run into this, but perhaps we don't properly emphasize some of these intangible benefits so that management can, at least subconsciously, translate them into profits for the company. Let me cite a half dozen which we recognize as important contributions from our standards program.

1. Promotion of Safety. Every standard is reviewed by our Central Safety & Fire Protection Division prior to publication. The standard thus reflects an approved installation or procedure. In 1961, our injury frequency rate was 0.39 per million man-hours of exposure, as compared with 3.71 reported for the whole chemical industry by the National Safety Council. We believe that standards have contributed to this record.

2. Resolution of Published Technical Data. Anyone who has been associated with chemical process development work is familiar with the number of sources for any one physical constant of a particular chemical. The question always arises as to which is the most reliable. As an example, for methylene chloride there are at least five authorities citing that many different critical temperatures. The capacity of a heat exchanger, the sizing of a distillation column, or the selection of an instrument, and, ultimately, the operation of a process may be affected by the temperature which is finally chosen. Our chemical process standards reflect the data which our best research engineers consider acceptable. This same approach is followed in many other standards fields such as acoustics, architecture, and power.

3. Uniform Engineering Practices. In today's expanding economy, new plants require an increase in engineering talent, both in design and construction. Currently, we subcontract a portion of our drafting requirements to handle peak design loads. Our standards are used by these subcontractors with the result that all drawings sent to the field reflect the required degree of uniformity. This practice undoubtedly helps reduce errors. A similar situation exists in construction. Our annual report for 1961 showed a construction expenditure of $205,000,000, and the engineering department's construction division handled the major part of this. Some 80 to 90 field subcontractors were involved, and standards played a big part in the execution of the contracts.

4. Criteria of Performance. Engineering standards are recognized as authoritative methods of performing repetitive engineering functions. As such, they may be used as criteria for measuring actual performance. Safety engineers use them in checking plant installations, inspectors for checking work performed by vendors. Finally, our manufacturing departments use them as acceptance criteria for a new production facility.

5. Training. With the present shortage of skilled mechanics, nearly all plants conduct training programs . . . the performance of a plant frequently depends on the success of its training work. Various sources of information are used, and it is a common practice to compile selected standards and use them in training for specific job assignments. Examples would be the training

of electricians in grounding principles or training welders in the handling of nonferrous metals or stainless steels.

6. Assistance in Early Plant Startup. When an appropriation is made for a new facility, it is everyone's responsibility to get the new plant into operation as soon as possible. This requires the use of every available means that will hasten the execution of the work; here again, standards become a vital engineering tool.

These are six intangibles we consider important. Your own program will no doubt suggest others, such as improved customer relations, or simplified servicing or repair.

## Summary

Sound evaluation procedures are necessary to justify our programs and to insure their continuation. Among the techniques which can be used are the following, none particularly novel, but all workable:

1. Develop a method for measuring the proper acceptance or need for a standard before it is issued. Your own figures on development cost versus average benefit expected will tell you how far you should go in this direction.
2. Establish a method for spot checking and evaluating individual standards. This not only insures that you really are getting the benefits you think you are, but also provides excellent publicity material to highlight your savings through standards.
3. Establish a system for measuring the tangible savings and the overall effectiveness of the standards program in dollars.
4. Recognize the intangible areas of evaluation.
5. Associate the standards program with other company cost-reduction efforts.

# HOW TO EVALUATE SAVINGS

Arnold M. Rosenwald

**Shirt-Sleeves Standards**

Dear Arnie,

Just got around to your column in the February issue of *Standards Engineering*. The subject of allocating standards charges is a very vital one and, as you point out, almost any approach has flaws. I believe we have considered every approach you mention. In general I believe that the standards expense is a relatively small one, so that allocation in many instances is not of earthshaking proportions.

What concerns me much more is the other subject you mention – evaluation of savings. It is fairly simple to get a standards operation under way. It's like motherhood – no one can be against it. So you get a token operation set up; everyone agrees it is producing results; and you want to expand to compound the advantages. You take your program to the financial and management people and they have to be very hard-nosed. How much does it save? There everyone bogs down.

There must be tried and satisfactory techniques for evaluating savings, but I haven't run into many as yet, neither in contact with individuals nor in the literature. DuPont's system is good, but somewhat limited in type of operation to which it may be applied.

Do you know of any good answers?

John B. Schafer
White Plains, New York

Dear John,

Thank you for your letter of May 15, in which you say in effect, "Never mind the cost distribution – how do you figure the savings?" In succinct reply to your last question, the answer is "No, I don't know of any good answers."

Now that we have dispensed with the only sure thing that I can say about this topic, please allow me to discourse on some of the things I think I know about standards evaluation.

First, the lack of ability in this area of evaluation is to my mind the single greatest deterrent to a flourishing of the standards movement in the United States. I place this item as the number one obstacle (even over the obstacle of selling management on the standardization concept) because, if we could evaluate better and more objectively, our sales job would be easier. (It would also be easier to enlist the support of the operating groups if they were made more fully aware of the direct benefits to them.) We would have more facts at our disposal; we could relate them better to product quality, performance and reliability – and we would thus demonstrate more ably that management support of a program is in the best interests of the company.

Taking the broad view, everyone feels that much can be saved through standardization. But when it comes down to answering the question, "How much can be saved in my company through standardization, and how much will it cost me?" the hard fact is that American industry doesn't know how to figure the answer. (Reference: ASA survey of 1959; *2,800* companies solicited, *238* replied, *34* quoted dollar figures on savings.) (Attention readers outside the USA: Do you have any savings figures?)

Second, I'd like to point out that I have avoided discussing standards evaluation in these columns because I didn't believe I had anything new to add to all that has been written in recent years on the subject. However, if your question is at all representative of those of the readers of these pages, then hopefully some purpose is served by my pulling together these thoughts.

Third, how else can we evaluate than to measure savings? We only standardize to save something – all of which things are translatable into labor or money or both. We standardize to save material, man and machine time, direct cash, etc. If we couldn't relate savings to time and money, where does the benefit lie? Hence, if we are to measure savings, we have to measure the chunks of time, materials, and money deducted from the productive process – be it manufacturing, intellectual effort, or other service performed.

This measurement demands that we know our costs both before and after the act of standardization. Now we are up against the nub of the problem – most accounting systems are not so sophisticated that all elements of cost are known. True, we can easily determine piece price by adding material costs, stop-watch times, and overhead, or by checking the vendors' invoices. But how do you accurately measure the design time to create the item?

We can easily figure the cost of a square foot of stockroom space by adding up the rent, heat, light, labor, etc., to maintain a square foot, but if we stock material on the spot and someone calls for it by a proprietary and unrecognizable name, how much does the special procurement action cost?

This next thought is a negative approach and really no defense, but it is also true that the very management people who question the justification for a standards program are at a loss to provide the cost data which are needed for proper evaluation. The standardizer thus put on the defensive is at a loss to provide the answers needed.

In this way, company standards ebb and flow with the tides of economy drives until something dramatic happens within the company to slam home to even the most severe critic that standardization, properly administered, is an essential management tool. Such dramatic events as unreliability in a mass-produced model or the mushrooming of a spare-parts program, worldwide, as a result of lack of parts standardization, have given rise to some large-scale standards programs in the past.

If you are looking for some specific areas for evaluation, I'd like you to look over a list developed last year during a brainstorming session in my office while we (Bud Blaher, Tom Daley, Bill Lyons and I) were trying to pinpoint ways in which standardization aided an organization in saving money. This list may include items which have appeared elsewhere, but, again, seeing them all in one place may be helpful to you.

I want to thank you for asking your question at just the time you did, because at the same time your letter arrived, my boss said to me, "Well, Arnie, it's time for this year's evaluation survey. Would you please develop one?" In answering your letter, I have decided to recommend presenting the topic list to our division controller and to solicit his conducting an objective survey. I'll let you know how it comes out.

Sincerely,

Arnie

**Profit-Makers Resulting from Company Standards**

*Inventory and Procurement*

- Reduce variety
- Reduce total
- Reduce record keeping
- Reduce storage area
- Reduce materials handling
- Larger quantities of fewer items
- Increased use of standards and decrease in specials
- More routine activity
- Reduction in P.O.'s, receipts, inspections, payments
- Reduced lead time
- Reduction in need for special tooling, training, layout, procurement, test

*Production*

- Familiarity in fabrication and assembly
- Less rework
- Reduced spares
- Reduction and simplification in inspection (sampling plans, etc.)
- Improved basis for vendor competition
- Improved quality control based on accepted and explicit specs
- Improved mechanization
- Economies of special purpose machines
- Reduced production time, industrial engineering effort and manpower

*Marketing*

- Ease of selling design composed of customer approved or recognized devices having reliability and maintainability

*Engineering*

- Greater reliability
- Reduced redesign, redrafting
- Ease of design based on standard parts and techniques
- Reduction in needed waivers and nonstandard part testing and approval
- Reuse of known items improves reliability and reduces debugging
- Reduces need for special communication between engineers, draftsmen, production, etc.
- Improved interchangeability of parts, designs, packages, tests
- Use of accepted standards versus untried specials
- Provide basis for built-in product improvement

*General*

- Reduction in special negotiations
- More accurate and predictable planning and budgeting
- Improved user and customer confidence
- Customer service – spares, manuals, operator training, repair, and training
- Provides basis for data mechanization, handling, reduction in errors
- Reduced capital requirement
- Improved general communication
- Provides basis for new bidding
- More routine work frees higher skilled people for unique aspect of project

P.S. If any readers have anything they would like to contribute on the matter of standards evaluation, why not send the data along to me? Maybe we can collectively improve our technique.

Arnie

# INDUSTRIAL STANDARDIZATION: A REQUISITE FOR SURVIVAL IN TIME OF NATIONAL EMERGENCY

Senator Ralph E. Flanders

Throughout World War II we had a higher degree of industrial standardization than any nation in the world. American mass production, made possible by the workable interchangeability of parts, was our Number One weapon in that war. Yet we cannot possibly estimate the loss we suffered in men and money, in time and resources, because of the lack of certain proper standards.

Our losses really began in the spring of 1940 when 400,000 Belgian troops might have fought better and longer if British ammunition had fitted their empty rifles.

At home we lost the services of thousands of small companies which would have participated in war production if there had been a comprehensive system of national defense standards to which they were accustomed to work.

Standardization is economically and socially desirable at any time, but in time of war or preparation for war it becomes nothing less than a requisite for survival.

How do we stand today? Both press and President have stressed the fact that we are in some ways, notably in industrial productive capacity, far ahead of where we were at the start of the last war. As far as standardization is concerned, that is remarkably true.

### Stock Numbers Surveyed

In 1945 the Navy – which had spent $55 billion during the war for the goods of industry – called in Remington Rand to help it survey its 50 million stock numbers. Of the first 15 million checked, only 2 million represented parts actually in use. Because some identical parts were listed under as many as 630 numbers, this 2 million in turn was reduced to 80,825. Today the project of simplifying and standardizing the stores of the entire military establishment is well under way.

A major effort is under way to resolve the differences between government practice and that of private industry in the making of engineering drawings. Requirements in radio and radar equipment are being standardized so that the military can obtain uniform and interchangeable products from many suppliers.

### Control of Standards

Standards do not come about by accident. They are born in a conscious effort; they are developed with organized machinery. Today the question of who is to control this machinery is a problem of our internal economy that is at once unrecognized, obscure, and crucial.

At the beginning of the industrial age our standards were written by only two men – the maker and the user. Early in the present century, however, the situation became involved with the growing complexities of our industrial society. The federal government joined producer and user as the third partner in the making of standards.

This intervention of government began as a means of curbing abuses in capitalism's iron age, of regulating standards affecting public life, health, and safety. Large government purchases in and between wars, all of them carrying specifications, next made federal agencies a directly interested party in many business and industrial fields. Centralized government buying according to purchase specifications began to cut deeper and deeper as a standardizing influence on industry. A half dozen federal standardizing agencies were created.

Government obviously has the right to set standards for the goods it buys; that much is granted. Yet there are trends, plans, and proposals currently under way that would make standardization wholly or mainly a function of government and I am opposed to them.

### Far-Reaching Effects

If you control an industry's standards, you control that industry – lock, stock, and ledger. On the day that standards become a governmental function and responsibility, as is now being threatened, the government will take a very long step toward the control of American industry. That is a step which will reach into every manufacturing and operating company, big and little, and consequently will affect every consumer in the country.

In such a setup, government personnel will decide when and what standards should be developed and what the provisions of the standards should be. That method is inflexible. It leaves little room for the manufacturer to experiment with products which do not comply with a standard. It does not allow for improvements in standards and their consequent revision. It does not permit the single manufacturer to depart from a standard in order to develop a specialized and useful business.

Standards made under such conditions tend to become limitations, controls, and restrictive procedures. They reduce consumer choice. The manufacturer obeys an order he had no part in formulating. Whether he thinks those orders are good or bad, right or wrong, he follows them, or he resorts to political pressures to get the standards he wants.

**Knowledge Lacking**

I am opposed to government control of standards activities, however, for a much simpler reason. No government planner knows enough to write the standards for the rest of American industry and all other American people. Even if he and his advisers were as wise as Jupiter, he would still stand helpless before economic forces which are too powerful, delicate, and complex for individual control.

Now, the situation certainly alters when people begin to shoot at us. In time of war we must quickly take special measures to save materials and manpower, to bring maximum efforts and resources to bear on production and mobilization. Under such pressure there are naturally tendencies and a will to give full authority to a planner or a group of planners, to hand down standards by decree.

A point to remember, however, is that they must still be effective, workable standards. The planner, of course, is no wiser simply because bullets are flying or because he has put on a uniform. The standards, the thousands upon thousands of purchase specifications they hand down, are as likely as before to be unrealistic, uncreative, and unnecessarily expensive, with this added difference – their errors can now be a thousand times more serious.

The solution to the problem lies in consensus. Where everyone is affected by standards, everyone should make the standards. Every group substantially concerned – manufacturer, designer, distributor, seller, consumer, and the government – should have the right to participate in deciding what the provisions of a standard shall be, and none should dominate the decision at the expense of another. For if a standard is to be of any value, it must be generally acceptable to all who use it. Industries, like people, are most willing to abide by regulations which they set up themselves.

The 30-year record demonstrates that voluntary standardization is an entirely workable and practical system. The American Standards Association, with headquarters in New York, is unique in this field and may be called private industry's answer to the challenge of government control of standards activities. It was founded in 1919 by five of the country's leading engineering societies and three Government departments – War, Navy, and Commerce – to bring order to the confused field of standardization. It now comprises 108 technical societies, trade associations and groups of such organizations, and some 2,200 private companies.

"If a standard is to be of any value, it must be generally acceptable to all who use it. Industries, like people, are most willing to abide by regulations which they set up themselves."

**Functions of ASA**

It works to simplify developments of engineering, commercial, consumer, and safety standards; to eliminate duplication, overlapping, and variables of standards activities by other standardization bodies in the country; to promote knowledge and use of standards; and to serve as a clearinghouse for information on all standards in the country. It serves as the channel of cooperation with the standards bodies of the twenty-nine other nations which belong to the International Organization for Standardization.

Since 1948 and particularly since the attack on the Korean Republic, the American Standards Association has been engaged in an increasing amount of work directly concerned with or related to the country's program for rearmament.

In working with the federal government for more than thirty years with a notable degree of harmony, it had already developed a formula for an effective cooperative relationship. As many as ten federal departments and agencies at one time were ASA members, and as many as sixty-four agencies have participated in the work of its sectional committees.

### The Directorate

Three directors and an associate director of the Bureau of Standards have sat on the board of directors of the American Standards Association, the Bureau itself helping to write nearly half of all American standards.

However, there is now a heightened danger that the intensified pressures of preparedness will cause the consensus principle in standards to be thrown out and war standards handed down by dictate.

There is a greater danger that the government, using the war emergency as an opportunity and an excuse, will not only take over full powers in standards activities but will fail to relinquish them when the emergency ends.

There is a group, both in and out of government, who feel that standardization is properly a government responsibility.

I think that neither now nor later can we afford such a reversal of our economic and technical processes. Our standards are now basic to our industrial economy. They will be more important than ever as we organize resistance against aggressive forces.

When this crisis ends, we must work to achieve a higher degree of harmony and order in our world; to relieve the strain of modern living by simplification; to increase the standard of living through more efficient production of interchangeable parts in a free market.

### Voluntary Agreements

I say that this is not work for the master planner. Creative dynamic standards are not composed on the higher levels and handed down by decree and proclamation. They are formulated by the voluntary agreement of all groups concerned, and must be worked out by the people themselves.

# INDUSTRY BENEFITS FROM STANDARDIZATION OF PRODUCTS AND PROCESSES

George T. French

The farm machinery industry is a very old one and my company has been busily engaged in it a long time – 128 years. Currently in our North American factories we manufacture farm tractors in 6 power sizes for a total of 38 basic models; we build about 450 different basic models of farm machines and implements, and for these we produce an almost infinite variety of optional sizes and extra attachments. In addition, for our industrial equipment product line we build about 50 different machines. For the manufacture of all this various equipment, we are required to make or buy 407,000 different parts and components. Our catalog of service parts lists 199,140 items and nearly all of these are kept available in our factory warehouses or parts depots. As might be expected, the raw materials required include a wide variety of products, used in a broad range of processes.

The finished product of the woolen mill is only raw material to the tailor, and the finished product of a steel mill is raw material in our operation. Let us therefore think for the moment of raw materials as products purchased, on which a considerable amount of additional work must be performed or a substantial amount of value added in order to build the end products which it is our business to make and sell. In the metal-working industry, raw materials covered by this definition have pig iron at one end of the list and paint at the other; and they include finished steel in its various forms – bars, plates, sheets, tubes, etc. – various nonferrous metals, plastics, fabrics, and a wide variety of components, such as bearings, drive chain, fasteners, and the various accessories for internal combustion engines. These are the primary building blocks which, through such processes as casting, machining, heat treating, welding, and assembly, are made into those end products which our industry sells.

In our standards program, which started over thirty years ago, the original objectives were to (1) provide means of identification of raw materials by class, type, and size; (2) establish specified quality levels and tolerances; and (3) provide for reducing the varieties of similar items. But the most important objective of standards, in my view, is to furnish basic data for communications between engineering and purchasing, between engineering and production, and most important of all, between the users' buyers and the suppliers' engineers and salesmen. Through the development of standards, American industry has created a reasonably precise language for buyer-seller communications about the properties of materials.

Other objectives have since appeared. The development of materials standards makes it more practical for suppliers to provide the same or related specifications to consumers in different industries, thus facilitating longer production runs. The steel supplier who can, in one production run, produce a material which will be used for farm tractors, motor trucks, machine tools, and road-building equipment has the opportunity of developing worthwhile economies.

Further improvements are only recently coming into the picture through the development of international materials standards. These broaden the range of purchasing activities, and also broaden the range of sales activities; we find they make it practical to manufacture U.S.-designed products in our plants in other countries, for example.

In the metal-working industry, typically about half of product cost is in raw material. In our company this figure varies from a minimum of 45 percent at one factory to a maximum at another of 63 percent. Clearly, with one-half the costs of manufacture involved in materials purchases, anything done to lower costs in this area will be beneficial competitionwise. Therefore, in pursuing its primary interest of maximizing profits, industry has turned to the standardization of raw materials almost as a universal first step in the control of costs. In our case we now have standards covering about 75 percent of our total volume of raw materials purchases.

**Savings**

In our company, the standards division justifies its existence by showing that its work produces savings which are much greater than its cost of doing business. Much of these savings are in the area of material cost reduction. As an illustration, our standards division recently developed a standard on "O" rings which made it possible to reduce the number of different rings used by 43 percent. This effected savings estimated conservatively at $50,000 annually. Somewhat to our surprise, a side benefit developed because the "O" ring standard reduced the varieties of back-up washers, thus creating a more favorable situation for competitive bidding and leading to further savings of $40,000 per year.

During 1950 our standards division introduced some medium carbon steel compositions having manganese ranges of 0.30–0.60 percent instead of the conventional 0.60–0.90 percent. These steels are used for applications that do not require heat treatment, so the lower manganese makes no appreciable difference in strength, but cost is 2 to 8 dollars less per ton, depending on the commodity. These steels we have designated as X1035, X1040, and X1045. Our use of these steels instead of the normal higher manganese steels, 1035, 1040, and 1045, yielded savings which last year reached $200,000. One of these steels now appears in listings of the SAE and AISI under a different designation, and many companies are today taking advantage of the lower cost.

Working with our materials engineering department, in cooperation with the engineers of our lubricant suppliers, our standards division developed a transmission oil for John Deere tractors which made it possible to reduce cost by over 20 cents per gallon, with no sacrifice in quality. Last year we bought approximately one million gallons of this oil, at estimated savings of over $200,000.

Paint specification is another area that points up the value of standards. In the United States we have developed some good standards and have worked closely with our suppliers. This is not yet true with our European factories and is probably the big reason that we have found the paint there to be lower in quality and at least 50 percent higher in cost. We are hopeful of introducing these standards in Europe, and have indications that they will definitely increase quality and lower costs.

In our business we use many fasteners – bolts, nuts, rivets, cap screws, etc. – and we manufacture a substantial quantity of these. When we first began to manufacture, there were no national or industry standards for materials for this purpose nor for the fasteners themselves. After long testing, we were able to develop a standard for four classes of bolts, separated according to physical requirements rather than by chemistry of raw material.

We believe that this was the first attempt in this field to build a standard predicated on functional requirements, and probably our standard was the original basis for the present SAE standard.

SAE now covers 9 classes, but for our own requirements we are still using only 4 of them. Notwithstanding, we found that we were still using too many varieties of size, types of head, and kinds of plating. Through the use of the SAE standard and a simplification program, we reduced the number of different items used from 25,000 to just under 4,000. This resulted in a one-time saving of approximately $500,000 and a substantial continuing annual benefit.

### Tangible Benefits Summarized

Standards have enabled us to define the quality of product we need and to avoid paying for something we do not need. Standards have likewise enabled us to make sure that our requirements are being uniformly met, thus reducing scrap and reclaim costs. We have been able to reduce inventory costs by reducing the varieties of similar materials and components, and, as a secondary result, standards have reduced the costs of inventory obsolescence. Reduction of variety has also enabled us to increase volume of purchased items, permitting our suppliers to effect and pass along economies through the larger volume. Standards have helped us simplify manufacturing processes and sometimes reduce the amount of special processing or separate handling which would otherwise have been required.

By enabling new vendors to compete for our business when they can meet the published standards, standards permit and even stimulate competition where it might otherwise not exist; this not only reduces costs but encourages progress. We believe that the development of our standard for paint, for example, has led to the development of paints with better covering power and longer life. As a further example, we believe that a new material such as ductile cast iron could have achieved such rapid and wide industrial acceptance only in a situation where standards already existed for the closely related products of malleable iron, pearlitic malleable, and cast steel. Clear standards have permitted potential users to compare specifications and have thus aided them to make informed decisions to use the new material. Certainly the existence of standards for man-made fibers and plastics has facilitated the development of new products in those categories. As industrial consumers, we sincerely believe that improved materials and better manufacturing techniques are natural results of better standards.

# STANDARDS AND THE FEDERAL CONSUMER

George W. Ritter

The federal government is the world's largest single consumer of goods and services. Its needs encompass the range of our engineering and productive capacity. It requires not only products of today's technology but things undeveloped which at this time are only ideas in the minds of men. To obtain the products to meet these needs, the requirements are pressed in terms of performance, design, construction, composition, quality, reliability, and others – prescribing them in various combinations. These requirements as they apply to many products are defined in industry or government standards. Unfortunately, this advantage is limited due to the absence of standards in many areas.

The breadth of my subject is such that it could only be treated in generalities within the time available. To enable me to deal with specifics, I am limiting my discussion to the use of standards as they apply to the areas of responsibility of the General Services Administration. A further reason for this limitation is that GSA is responsible in areas of common interest to the individual consumer.

The role of the General Services Administration in government supply operations is to develop and maintain jointly with the Department of Defense an efficient and economical national supply system for federal agencies. To this end GSA develops, implements, and monitors on a continuing basis and in close coordination with the Department of Defense procurement and supply policies and methods governing the procurement and supply operations of executive agencies. An integral part of this mission is the operation of various programs including procurement, distribution, quality control, cataloging, and the development of specifications and standards.

In its capacity as the supplier of the so-called "commercial type" items to government agencies, GSA's procurement exceeded $1.5 billion during the last fiscal year.

### Single Federal Catalog

I mentioned cataloging as a part of GSA's mission. You may recall that a government-wide program was initiated a number of years ago to establish a single catalog system for the federal government. This system was designed to identify each item procured, stocked, and distributed by any government agency. Techniques were developed to reveal like items and to differentiate unlike items.

The Department of Defense completed the cataloging of military supply items in the federal catalog system in 1956. The program for the cataloging of the items in the civil agencies of the government was scheduled after the completion of the Department of Defense program. I am very happy to report that the active items used by the civil agencies will be cataloged in the federal catalog system very soon.

For the first time in our history there will be a record of the actually different items which federal agencies procure regularly. This record reveals their similarities and differences and whether or not they conform to established standards. It provides basic data for technical analyses permitting the elimination of unnecessary varieties and the establishment of standards.

I have mentioned GSA's role in government supply operations. Let me elaborate briefly upon that role as it applies to standardization in the federal government.

### Objectives of Federal Standardization

The General Services Administration prescribes policies and procedures for the development and use of federal standards and federal specifications by government agencies. They are developed either by the General Services Administration or by other government agencies by agreement. The Department of Defense has 39 percent of the active development projects,

GSA 52 percent, and other government agencies 9 percent. In all cases they are developed in coordination with industry and with interested government agencies. This procedure is essential to correlate the needs of the government with current technology and industry practices. Federal standards are developed to achieve the following objectives:

Reduction in the number of sizes, types, and grades of items used by federal agencies and use of the minimum types of packaging consistent with economy and efficiency of operations

Insuring purchase of only such items which are adequate to meet essential requirements of federal agencies but which do not unreasonably exceed those requirements

Attainment of the greatest practicable degree of uniformity of dimensions or other characteristics which will facilitate interchangeability of parts and replacement of components

Achievement of the highest practicable degree of uniformity in the practices and processes essential to the manufacture, preservation, packaging, packing, inspection, and acceptance of items for the purpose of achieving maximum economy in these operations

Adoption of uniform terminology and definition of technical engineering, and supply practices and processes

Utilization where practicable of nationally recognized federal agency, industry, technical society, and other commercial standards

Selection as standard of those products which are of standard commerical construction, manufactured by business enterprises for general commercial use, consistent with essential needs of federal agencies

Concentration of standardization efforts on common-use items having large purchase volume and which are of importance in the supply programs of federal agencies, thus permitting the greatest savings to the government

Conservation of critical materials to the maximum extent practicable

From these objectives, you can see that federal standards cover a wide technical field. They establish engineering or technical limitations and applications for materials, processes, methods, designs, drafting room, and other engineering practices. They are designed to achieve the highest practical degree of uniformity in materials or products or interchangeability of parts used in those products.

Now, let us look at this from the viewpoint of federal agencies as consumers. What do they get out of this standardization effort?

The benefits realized from the use of federal standards are revealed tangibly in direct dollar savings in procurement and other supply operations. In addition, and of no less importance, are intangible benefits resulting from the use of uniform engineering practices, terminology test methods, identification markings – all of which contribute to more effective interagency and government-industry relationships. Other standards designed to protect the health and provide safety to individuals demonstrate their value through the reduction in human suffering.

To demonstrate the potential benefits to the government as a consumer, which are derived from standardization, I have selected a few random examples in several areas in which standards have been developed.

Before a standard was developed for felt-tip markers, we bought 32 items. After standardization, 8 items satisfactorily meet the needs of federal agencies, at a savings of $162,000 annually.

A federal standard now prescribes 34 fabric typewriter ribbons and 2 teletypewriter ribbons. Before its development, 68 typewriter and 62 teletypewriter ribbons were purchased at an annual cost of $150,000 more than the standard items now cost.

Standards in the field of packaging reveal an exceptionally fertile field for savings. One standard developed for bulk packaging or masking tape is conserving $247,000 per year compared to the cost prior to its establishment.

Federal Standard 183A prescribes uniform identification marking over the entire length of iron and steel products such as bars, sheets, plates, and tubing. This standard is referenced in 65 federal and military specifications.

Standard test methods enable government and industry laboratories to evaluate products on a comparable basis. This minimizes the production of products by industry which are unacceptable by the government with the consequent contractual delinquencies and economic hardships.

The "payoff" from standards designed to protect the health and safety of individuals can only be assessed by the extent of their inadequacy.

A considerable amount of attention is given to the establishment of standards which will prevent dangerous situations from occurring. Here are some of the examples.

Standards have been established for safer tools used to drive fasteners and anchors into steel and masonry materials.

Standards for dry chemical fire extinguishers not only provide for superior performance over carbon tetrachloride extinguishers, but they remove the hazard to personnel through exposure to toxic vapors.

Undoubtedly of great interest among our standards for the protection of personnel are our proposed standards for automotive passenger safety.

Public Law 88-515, approved last August, prohibits the purchase by the federal government, after the prescribed effective date, of any passenger automobile which does not conform to certain passenger safety standards. The Administrator of General Services was directed to prescribe such standards within 1 year from the date of the act to be effective 1 year and 90 days thereafter.

Seventeen proposed standards have been developed with the assistance of organizations within the government and industry which are concerned with passenger safety in automobiles. The proposed standards were published in the *Federal Register* of January 26, 1965. Comments are due by February 26. The official standards are scheduled for publication on June 30, 1965. We recognize that these standards are but the initial effort. Those published will require upgrading and new ones must be developed. It is a continuing effort.

The examples which I have just given are representative of the many benefits which government agencies derive from the use of effective standards. Based upon dollar savings alone which result from our standardization efforts we have computed a savings to the government of $50 for each dollar spent.

### Benefit of Federal Standards to the Consumer

I have limited my discussion to the benefits which federal agencies realize from the use of standards. I do not wish to leave the impression that the benefits from the government's use of standards are confined to the government. As I mentioned earlier, GSA is interested primarily in the same types of items of interest to the individual. The standards developed for such items for the government are frequently highly acceptable to the private consumer. There are many instances in which items are advertised in the trade as conforming to government standards. In such cases, the consumer has a basis upon which he can evaluate the product he buys.

In the case of the safety standards for automobile passengers, it is highly probable that many states will adopt some or all for vehicles which they license.

In conclusion, valid proof supports the value of standards – be they government or industry – in the day-to-day operations of government and in the daily lives of individuals. To realize increased national benefits from the great potential which standards provide, both government and industry must demonstrate a more active interest and provide resources necessary to maintain a dynamic standards effort.

# Part 5

# COMPONENT SELECTION AND PARTS MANUAL

*Helping designers with finding a component for a project, evaluating components for the purpose of standardization, i.e., including it in a standard parts manual, is a routine job for standards and components engineers. "Panel Discussion on rotating servo-components" will take you to the heart of this subject; it will clarify not only what must be done in order to find a desirable component for a specific condition of use, but also how much experience with the components' application and with standards and specifications covering the components and know-how in general are necessary in order to make a meaningful, successful choice of a component.*

*R.G.*

# PANEL DISCUSSION ON ROTATING SERVO-COMPONENTS

Presented at the Thirteenth Annual Convention of the Standards Engineers Society, September 2, 1964, Hotel Commodore, New York City

**Panel Members**

Marvin Walters Head, Components Branch, Engineering Division, Bureau of Naval Weapons, Washington, D. C.

Benjamin Sachs Director of Reliability and Quality Assurance, Vernitron Corporation, Farmingdale and Long Island City, New York, and Torrance, California

Albert Diamond General Manager, Farmingdale Division, Vernitron Corporation, Farmingdale, Long Island, New York

**Moderator**

Ralph Pickus Supervisor, Components Group, Sperry Gyroscope Company, Great Neck, New York

*Mr. Pickus:* We will start with questions for Mr. Walters, our government representative.

The first one – and this has to do with synchros – is, Let us suppose you need a synchro in your new design of a weapons system. How would you go about finding which synchro to use, where to purchase such a synchro, and how to install it?

*Mr. Walters:* In taking this question I assume, first, you may not know where to find a synchro, so we start with the fundamentals of going to the *Department of Defense Index of Specifications and Standards*. Then you look up the word "Synchros," and you will find a general specification quoted, which is MIL-S-20708. You will also find a standard quoted: MIL Standard 710.

Using those two documents and reviewing MIL Standard 710, you will learn there are 72 types of standard synchros recommended by the government. We divide these into torque synchros and control synchros. It is the responsibility now of the designer of the weapons system to choose the right size of synchro, the right primary voltage, primary current, primary power, impedance angle, transformation ratio, phase shift, electrical error, friction torque, radial play, end play, torque gradient, temperature rise.

Now, each synchro is identified by a type number, such as 26V-11TX4b. If you go back to the general specification MIL-S-20708, you will discover this type designation has a special significance. The 26V stands for 26 volts. The 11 refers to the diameter to the nearest tenth of an inch. The TX refers to torque transmitter, the 4 refers to 400 cycles, and the letter b refers to the modification of the design.

A commercial synchro that does not meet the military requirements of MIL-S-20708a for physical dimensions or shaft dimensions or electrical parameters is forbidden to use a military standard type designation.

Such synchros are given commercial designations by each manufacturer. When you go to your detailed specification – and in this particular case it's /6A with the general specification preceding it, like MIL-S-20708/6A – you will have your detailed performance, the outline dimensions, the shaft dimensions.

In the event, now, that a synchro type is not included in the MIL Standard 710 (the 72 types that I mentioned), and you require a special synchro, then you should obtain approval from the contracting agency for using a special synchro.

We discourage special synchros as much as we can, and this is spelled out in rather strong language in our MIL Standard 710.

The next question is, where do you purchase such a standard military synchro?

As you review your detailed specifications you will notice on the requirement a statement which says that "The synchro furnished under this specification shall be a product which has been tested, passed the qualification test specified herein, and has been listed on or approved for listing on the applicable Qualified Products List."

This list giving access to this information is available throughout the country from the various military offices in the United States; also the Bureau of Naval Weapons keeps an up-to-date file of the QPL lists. We also have it at our Synchro Test Facility, Naval Ammunition Depot, Crane, Indiana. When carried to the full extent, we will have 72 QPL lists. But each such list gives the names of the synchro manufacturers, the addresses of the synchro manufacturers, and only those that have passed the Qualified Products Test, the QPL test.

We speak about the installation of a synchro. This is a rather simple matter. We use either a splitring that fits into the groove in front of the synchro, or we use 3 or 4 so-called synchro clamps. These clamps are evenly spaced, and they are bolted from the mounting flange of the instrument. Installation drawings show that clamping arrangement. These drawings are referred to in the detailed specification.

I might mention that there are other publications that are valuable in using synchros. We have recently put out a new military standard handbook, *MIL Handbook 214 (Weapons)*. This provides engineering and technical data on the standard synchros and those that were built for earlier-existing systems. Also, there is a very old publication which we are bringing up-to-date. It's OP 1303, which gives descriptions and operations of synchros, including the hardware, electrical connections, loading information, and other valuable information on how to use synchros. These last two documents can be obtained from the Bureau of Naval Weapons or the U.S. Government Printing Office.

*Mr. Pickus:* Is the procurement of MIL type synchros the responsibility of the systems manufacturer, or are they supplied as GFE (government-furnished equipment)?

*Mr. Walters:* Today, synchros are the responsibility of the systems manufacturer – to procure his own. The government is no longer supplying these as government-furnished equipment. This policy was changed about two years ago. Several reasons for this are that this now allows one responsibility for procuring the components for the system and meeting production dates and keeping the schedule according to an overall program.

In order to help the manufacturers or the systems designers and contractors to procure the proper synchros, the Bureau of Naval Weapons, starting about two years ago, started the so-called Qualified Products Lists. We did not have those until two years ago. Before two years ago we were procuring these synchros as GFE equipment. We were testing them at Crane, Indiana, we were stocking them on the shelf, and issuing them for research purposes, for production systems, and for new designs. So today, having changed this policy, we considered it necessary to establish a list of two or more available qualified manufacturers who could make a standard military synchro to our specifications.

Now, going through this procedure, we found that there are certain rules that we had to meet before we could set up a qualified products list, and I might briefly mention the five steps that we had to go through.

First we had to have a general specification up to date, which required the qualification tests to be in force. So we rewrote our MIL-S-20708, revision A, and put it into a revision B, and called it a limited co-ordination document, as a BUWEPS document.

Secondly, we had to back up that general specification with an up-to-date detailed specification, which we did by rewriting the detailed specifications. At this time, for the first time, we introduced impedance values for rotor and stator windings. We had to do this in order to standardize our product still further than we had done.

Then we had to announce, through the *Department of Commerce Business Daily* and mailing lists and trade journals, the fact that we were going to establish a QPL list on synchros.

Four, we had to have some test facility running to evaluate the manufacturers' products. This test facility has been one we have had for a number of years at Crane, Indiana.

Then we had to set up a way of reporting the test results, the permanent filing of the test results, and the publication of the lists.

All of these things we have complied with, and today we have established approximately 12 separate QPL lists. We will take this up to the number of 28 types in our present program, which will carry us through probably another one year of testing.

*Mr. Pickus:* The next question is, What criteria have been established to determine acceptance or failure of synchros being tested for QPL approval?

*Mr. Walters:* When we were testing our groups of synchros we realized we could not use just human judgment and emotional feelings for grading these different manufacturers who had submitted their products. So with several meetings between our test laboratory and the Bureau of Naval Weapons, we established a point system. This point system involves the four samples that are submitted by each manufacturer for his test.

It graded out such that if a manufacturer was perfect in meeting all the specification requirements, he could achieve 560 points on his four samples. We made the passing grade 492 points or approximately 88 percent. The number of points that we assigned to the 7 parameters were weighted in favor of those parameters which we considered the most important.

For example, electrical accuracy carries the highest number of points in each test group. The 7 parameters are brush contact resistance, radial play, end play, high potential, insulation resistance, accuracy, and friction torque.

Now each of these 7 parameters is checked 8 times during the cycle of the test. These 8 cycles are, first, when we receive the units at our test facilities; they are checked in just as they come from the manufacturer. We test them again after we have vibrated the samples, then after we have subjected them to 50G shock tests. Then we subject them to a low temperature of -55 degrees C, and a high temperature of 125 degrees C. Then we give them the humidity test, the endurance test, and, finally, the 2,000 foot-pound shock test.

A total of 80 points per sample can be scored by meeting the accuracy requirement. Brush contact resistance permits scoring only 13 points per sample. The completion of the endurance test is heavily weighted to permit scoring a total of 40 points per sample. Two samples must endure 500 hours each, or the entire lot of 4 samples must be resubmitted.

*Mr. Pickus:* What has been the most significant deficiency revealed by the QPL testing, and what has been done about it?

*Mr. Walters:* We discovered during our QPL testing (and realize that we had roughly ten manufacturers submitting their products) that we were not passing the endurance test, and this was true for more than one manufacturer. In fact, the majority of our manufacturers were not passing the test, the endurance test, which is 1,200 hours of operation at 1,150 revolutions per minute at various temperatures and with the shaft in various positions.

When we learned that these units were failing after 200 to 300 to

400 hours, we knew we had a problem there; in fact, it was probably testing a large group of different manufacturers' samples submitted under the same specification at one time that gave the best test result we ever had in synchros.

We found that the synchros were binding up and not turning. Looking into the bearings, we found a reddish rust-like stain. We found no lubrication evident. The synchro bearings were dry. Fortunately, the Bureau of Naval Weapons has access to a special group of lubricant experts. They have devoted years to the study and research of lubricant problems.

We called this team in to look at the problem. We also wrote to each of the manufacturers to ask them what type of lubrication they had used in submitting samples. We found that in the majority of cases they had used just a few drops of silicone oils.

You realize that a synchro must operate at ambient temperatures of -55 degrees C to 125 degrees C. At the lowest temperatures there is a minimum torque requirement, and at the high temperature the lubricant must not break down. Also, shelf life of the lubricant is important. There was much mystery about the torque values of different lubricants at the low temperature.

Our Naval Research Laboratory set up a torque measuring device at different temperatures and started to measure different types of lubricants to see which torque specifications could be met with different lubricants.

We investigated the silicones, diester oils, diester greases, and combinations of the silicones and diesters to arrive at acceptable torque values. Finally, the Naval Research Laboratory recommended certain types of oils and greases for some endurance testing at Crane, Indiana. Also the Naval Research Laboratory had discovered a few years ago a chemical barrier coating that could be applied to steel parts which would prevent the migration of lubricants. These were also applied to certain parts of the ball-bearing area to contain the lubricant within the ball bearing at the place that we need it.

After we put these barrier coatings on a group of synchros at the Crane, Indiana, plant, we put them through the endurance test. We had three groups of Size 11 synchro, five synchros in each group, so lubricated and treated. Using different types of lubricants and the barrier coating, the results that we obtained were very gratifying.

The Naval Ammunition Depot conducted these tests using special test fixtures. They made measurements of torque at low ambient temperatures of -55 degrees C. The tests were made with temperatures elevated to

135 degrees C ambient. This was to be sure that our lubricant would not break down in a new test.

The findings that we achieved we summarized briefly as follows:

Those synchros lubricated with ample quantity of diester oils and greases and employing the barrier coatings have exceeded 2,500 hours of operation. Some are still accumulating hours. Those synchros lubricated with a few drops of silicone oil, or a few drops of mixed lubricants and with the barrier coatings, failed to pass the endurance test.

It is therefore concluded that synchro reliability can be improved by at least a factor of 8 through the use of government-recommended systems of lubricants and the application of chemical barrier films to prevent lubricant migration.

The migration occurs most severely during the vibration tests.

The experiments and tests are continuing on synchros, and we have a similar program under way for servomotors.

*Mr. Pickus:* Thank you very much, Mr. Walters. We have a number of other questions for you, but in order to spread the time around, I beg your indulgence and I'd like to go over to Mr. Sachs.

Mr. Sachs, we have here Question No. 1, something that I think everybody is interested in. The question is, How much does it cost to verify an MTBF, and how can you do it?

*Mr. Sachs:* Taking a very straightforward approach, if we assume that there are sufficient components available to simply run failure-rate tests on them, we can very easily determine what the cost of such a program would be. For example, if we were interested in verifying a mean time between failures of $10^4$, or 10,000 hours, which corresponds to a failure rate of 10 percent per 1,000 hours at a confidence level of 60 percent, we would need in the neighborhood of 9,300 hours of test data.

Now, if we assumed that we were dealing with a component costing about $50, and we could afford to neglect the cost of the testing and just consider the cost of components to be involved, and we were going to assume a 1,000 hour operating life of the component, that program for a $10^4$ mean time between failures would cost in the neighborhood of $500 for a $50 component, which is well within reason.

If we needed a higher confidence level, say a 90 percent confidence level, we would have to multiply the number of test hours and therefore the number of test units by approximately 2½. This is for a minimum program, in other words, a program in which no failures are going to be permitted.

In the event that you experience a failure during this period of test time, then all bets are off, because you have to go to a different type of statistical design where you permit failures; and therefore you need many, many more test hours in order to verify your mean time between failures. A mean time between failures of $10^4$ is a relatively modest requirement. We are seeing requirements these days of $10^6$ and even $10^8$ or $10^9$.

Using the same $50 component and 60 percent confidence level, to verify a mean time between failures of $10^8$ or 100 million hours, would cost $5 million just for the components to be tested.

For a 90 percent confidence level, we multiply that by 2½, and we are talking about $12 million. This again is for a minimum program involving no failures occurring during the entire test time.

Just to give you a feeling for what happens if you want to set up a program where you are going to allow one or two failures for the 60 percent confidence level: for a program to permit one observed failure you would multiply the test hours, and consequently the expense, by two. For a program to allow two observed failures, you would multiply by three. The multipliers are considerably higher if you want to go to a 90 percent confidence level. What that means, really, is that when dealing with a relatively low volume item, like a rotating component, unless you are looking for an extremely modest mean time between failures, it is economically not feasible to have an empirical demonstration of your reliability. This has been recognized pretty much by the systems producers in recent years, and most of the specifications that we have seen in the last few years that called for very high reliability rotating components have specifically indicated that an analytic, rather than an empirical, demonstration is required.

The reason, very clearly, is that an empirical demonstration of extremely high reliability on rotating components would involve a quantity of the same order of magnitude as all of the rotating components produced in the country each year.

Now, when we consider an analytic approach to a reliability demonstration, we can begin with the presently approved *MIL Handbook 217* method. In this *MIL Handbook 217* they have divided up the failure rate of rotating components into a portion derived from the windings and insulation failure possibilities, which is analogous to the method given for transformers, and they have added to this failure rate for the windings an additional failure rate for the bearings and brush to ring contact. The *MIL Handbook 217* approach has the advantage of being rather straightforward and inexpensive.

The difficulty with the *MIL Handbook 217* is that it tends to lump all synchros, all servomotors, and so forth, into one category, and completely eliminates from consideration a lot of the very significant differences that exist between these components.

For example, we have in the neighborhood of 100 slash specifications to MIL-S-20708, meaning that there are perhaps a hundred MIL type synchros involved, and a large proportion of these are fairly active. Some of these synchros are wound with very fine wire, some of them are wound with relatively coarse wire. In some of them the coils are packed very tightly into the stator stack; in some of them the coils are loosely packed. This difference in wire size and in packing contributes a difference in the mean time between failures in these components that is much greater than any of the factors that are taken into consideration in *MIL Handbook 217*. Using the rather unsophisticated *MIL Handbook 217* approach, you would completely lose all of those distinctions.

By the same token, there are fairly substantial differences between bearings and between different types of brush and collector ring constructions. Again, these differences are simply not taken into consideration if you apply straightforwardly the *MIL Handbook 217* approach.

Consequently, we have come to the conclusion that in order to come up with an analytic demonstration of a mean time between failures for a rotating component, it becomes necessary to analyze the specific construction features of that particular component, rather than to lump all synchros together, all servomotors together, all motor generators together, and so forth.

Another very important consideration is the application of the component in the environment in which it is going to be used. In certain rotating components, the type of switching that they will be exposed to, the type of transients that they will encounter, becomes very important, so that our answer to the question of verifying MTBFs is that unless you are looking for an extremely modest type of MTBF, it becomes necessary to consider your particular component in the light of your particular application and your particular environment.

That covers, I think, the design reliability aspect of the situation. When we talk about reliability, we have to recognize that we are talking about an aspect of statistics that goes very far beyond what we have encountered in quality control work in the past.

We have commonly been subjected in the industry to government-imposed AQLs, or acceptable quality levels, of one percent, meaning that the sampling plan was designed quality-controlwise to accept lots which were one percent defective. These are the ground rules that prevailed up until a few years ago with respect to quality levels that were considered acceptable.

When we talk about very high reliability it becomes absurd to also talk about a quality level that is going to permit one percent of the units to be defective, because that one percent would completely obliterate the very, very minute proportion of reliability failures that we theoretically could tolerate.

So that, along with any design program, we must, in the production of components, enforce a type of quality control and traceability control over the materials, the processes, and the inspection procedures that goes very far beyond what was considered normal practice for good MIL-type components only a few years previously.

To try to give a specific answer to the question of cost, I have come up with some figures which represent what our company has actually been quoting for different reliability levels. These costs include, in general, a reliability design effort that is necessary on the component, and the manufacturing traceability and control effort that we feel is required on high reliability components, beyond what you would normally expect on a good MIL-type rotating component.

These figures are as follows: For what we would term high reliability, that is, a mean time between failures in the range of $10^4$ to $10^5$ hours, we have been quoting up to about 20 percent above the basic unit price for reasonable quantities of the component. That would be quantities of 50 or more.

For what we would term very high reliability, that would be mean time between failures of $10^5$ to $10^6$ hours, we have been quoting in the area of 20 to 50 percent above the basic unit price, again in quantities of 50 or more.

And then, getting into the area of ultra high reliability components, having mean time between failures of $10^6$ up to about $10^8$, we have been quoting from 50 to 100 percent above the basic unit price.

There is quite a range there, because the complexity of the designs differs, and the basic price might differ. Generally speaking, we find that the ultra high reliability requirements tend to come up on nonstandard items, so that the cost of the item is usually not in the $50 range that I mentioned earlier, but more likely the $350 to $500 range.

*Mr. Pickus:* Question No. 2 is: How does a contractor purchase a high reliability MIL part, or is the terminology conflicting in terms of the standardizing procedures and the way the specs are written?

*Mr. Sachs:* MIL-S-20708b provides certain reliability levels. The lowest level that is provided is a MTBF level of about 33,000 hours, or 3 percent per 1,000 hours failure rate. The other three levels that are provided are 1 percent, 0.1 percent, and 0.01 percent per 1,000 hours failure rate.

On the basis of published figures on rotating components, the 33,000-hour mean time between failures is a fairly moderate, achievable figure, and any recognized supplier of MIL-type rotating components should be able to guarantee at least that level.

When you go beyond that level, you are entering the area of reliability verification where I think the customer reasonably has to be prepared to pay for that verification.

*Mr. Pickus:* The next question is: Could you recommend a reliability type of test on possibly a sample basis or maybe even 100 percent, which would stress the unit but not impair its life significantly, in such a fashion that those units that passed would absolutely (and more than is now likely) pass their design life, and those that didn't would be culled out? This is for motors as well as synchros.

*Mr. Sachs:* The one thing that can be stated with regard to this question immediately is that there is certainly no overall answer to this question. The failure mechanisms of rotating components are too numerous, too varying, and too complex to come up with one simple burn-in type of test that would work for all of them.

For example, at one time it was considered very desirable to use a higher voltage high potential test than is used today. It was believed that you had a certain proportion of weak sisters in your lot of components, and if you used a high voltage high potential test, you would weed out these weak sisters and the remaining units would then represent a higher reliability group of units.

We have run some very extensive testing on this theory, and have proven it to be completely false. Generally speaking, the units which have survived a series of high pot tests at high voltage are in poorer condition than when they started out, and for this reason the practice of using these high voltage tests has pretty well been done away with and at the present time the specs call for reducing the high potential voltage after the first test has been run.

Certain military specifications for high reliability capacitors provide that if you apply a 40 percent over-voltage you achieve a 500 percent acceleration factor, and you can therefore reduce the number of units to be failure-rate tested by a factor of 5.

Now, it would be very fine for us if we could find a simple approach like that that would work on rotating components; however, we have the failure mechanisms on rotating components that operate on bearings, those that operate on the collector ring and brush contact, those that operate on winding insulation, those that operate on connections, and so forth, and the importance of each of these failure mechanisms varies very substantially between components. A component that is wound with fine wire is more susceptible to one type of failure mechanism. A component that operates at high speed is susceptible to a different type of failure mechanism. The only answer that we can give to a question like this is that you can probably work out a burn-in test procedure for any component, but it has to be done with the assistance of the producer's engineers and reliability personnel, and it has to be done in consideration of the particular application and environment for that component.

*Mr. Pickus:* I will now direct the engineering questions to Mr. Diamond.

The first question is: How can I avoid "designed in" unreliability when specifying a rotating component?

*Mr. Diamond:* Basically, the answer to Question 1 is that you do not write or design reliability into an existing component, so do not put reliability into the specification as an afterthought. Work it in with the intrinsic or basic design of the unit. Too many times we find that here we have, let's say, an example of a simple device we all understand, a motor. This motor is burning hot. It's been designed to operate into a system which requires that a miniature component put out a torque far in excess of what is normally expected.

Then the question is asked: How can I make this unit reliable? Well, common sense would tell you that if you put the proverbial egg on it and it fries the egg, that this unit intrinsically is not reliable. So reliability should have been started at the concept stage, rather than as an afterthought.

*Mr. Pickus:* You are seeing to that beforehand.

*Mr. Diamond:* I am. The concept, actually, is going to have to be part of the basic design.

*Mr. Pickus:* The second question is with regard to miniature synchros and resolvers (size 8 or smaller): Does increasing the impedance adversely affect accuracy and reliability?

*Mr. Diamond:* In answer to the first part of the question, we have to understand what affects accuracy in a synchro. Outside of the normal factors of distributing your winding, two major factors will affect the accuracy. One would be the source impedance and then the air gap, or concentricity of the parts. In this case, anything which affects these two factors will be adversely affecting accuracy.

Increasing the source impedance affects accuracy; therefore increasing the impedance of the unit, which necessitates wire size changes, which increases the source impedance, will decrease the accuracy.

Furthermore, increasing the impedance so that it is required to decrease the air gap will cause a falloff in accuracy. Notice that the question was with respect to miniature synchros.

Now, the effect is more noticeable in miniature synchros because in your larger synchros you have enough area to increase the impedance without compromising on the pack of the unit, so up to certain limits you are able to increase the impedance of the large units without adversely affecting accuracy.

But even in this instance there is a limit, and it's all relative. Once you are forced to go to a wire size which increases the source impedance, or you are forced to decrease the air gap, you will then adversely affect the accuracy.

Now, in terms of the reliability aspects of it, we look at these two factors: Obviously, decreasing the air gap, whether it be a motor or synchro, must mean that the unit is less reliable because you have more chance of rub failures or striking failures, or if the bearing were to wear slightly, you get an increased error of the eccentricities involved. So therefore, again the answer is Yes, increasing the impedance will force a decrease in accuracy and a reliability factor which is worse than that in a unit which has a more realistic impedance level.

*Mr. Pickus:* The next question is: How should resolver accuracy be specified to assure that system requirements will be achieved without excessive costs being incurred?

*Mr. Diamond:* Ground rule: The ground rule is, specify the resolver the way you are going to use it. Don't necessarily lean upon existing specifications or standards which are in your company. By this I mean the following: Let's say that it is according to MIL spec to check a resolver by checking axis error and checking conformity to assigned function on either winding.

Let's assume you are going to use this in a transmission system as a control transmitter. Well, if you wanted a 7-minute control transmitter accuracy, by using the extension of logic which says that I need 0.05 percent conformity because I have two windings to worry about, and a 3-minute interaxis error, and if all the errors were on the extreme limit, I would end up with a 7-minute unit, means you are specifying a unit which in general is far more accurate than you need.

You should specify it as a 7-minute CX, tested as a 7-minute CX, so you do not pay for the additional quality cited previously. There are also other tricks to achieve accuracy, and that is skewing the excited member, which would completely defeat your purpose of tests, were it to be rotor excited in tests.

These are all the parameters you have to worry about, really, and the ground rule, again, is specify the resolver the way you are going to use it. Do not use any subterfuge in the tests, or specious reasoning involving the testing of the resolver.

*Mr. Pickus:* Generally, a motor-generator combination is used in an all-AC servoloop for damping purposes. Can other damping methods be employed that would increase reliability and reduce cost, size, and weight?

*Mr. Diamond:* The way this question is worded, I think that what is meant is that in all-AC servosystems one of the conventional methods used for damping is the motor tach or motor generator. There are other methods available for achieving similar results, and some of them give superior results to motor generators. One of them is the adjustable velocity damp, which is the electromagnetic equivalent of rate feedback. These do have certain disadvantages, such as the fact that you absorb more energy in the motor damping function. However, there are many instances where this can directly take the place of a motor generator with increased reliability and reduced cost, since there is reduced waste. Also, you save a channel in your amplifier.

There are other damping methods used in all-AC servosystems. In the electronics area it is called a notching filter. The equivalent of this is called an inertia damp. This gives correction for stability along with no reduction or a reasonably small reduction of velocity.

*Mr. Pickus:* Question No. 5 is: Specification of an integrating tachometer (temperature compensated) is particularly involved. What parameters are most critical in regard to system's error contribution? How should they be specified? What are figures for these parameters that are practical?

*Mr. Diamond:* First of all, we concentrate on one parameter called linearity. The linearity of a normal type of integrating tach is in the order of magnitude of, let's say, 0.07 percent. The change in gradient versus temperature of an uncompensated integrating tachometer, that is, before you add your compensation, is in the order of magnitude of 0.05 percent per degree Centigrade. This means that a 2-degree difference in temperature would already take the gradient change over the permissible nonlinearity of the unit. One would be 0.1 percent as against a 0.07 percent. Therefore we should really concentrate more heavily on specifying the temperature gradient variation, rather than the linearity specification.

The other area of concentration would be in thermal parameters. We should specify axis error and axis change over the temperature range. We should specify the position error as a function of room temperature and temperature range.

However, contrary to common beliefs, there is very little position error change versus temperature. The major change is on the axis. More concentration should be placed upon the characteristic of the motor end of the motor generator, since it is extremely important to be able to integrate down to low velocity. Therefore, some specifications relating to maximum cogging or variations of positions of the rotor for input voltage should be specified.

*Mr. Pickus:* Right now I'd like to go back and ask Mr. Walters some additional questions.

Is a MIL unit only purchaseable by a government agency? In particular, where there is no QPL, what prevents any vendor from using a government nameplate?

*Mr. Walters:* When we say, Is a MIL unit purchaseable, we know that for existing systems we do have to buy units that are not the MIL standard synchro unit. These are commercial type units.

However, we are making an effort each day to convert these to the MIL standard unit by checking the parameters back and forth, but we know that for our so-called stocking spares, we do purchase units that are not the MIL standard.

How do we prevent a synchro manufacturer from using a government nameplate? There is no way. We do not have a police force for this. However, we feel that our reputable firms which are listed on our QPL list today are not doing this and would not stoop to do this. We feel our QPL list is an indication of where we would expect high quality and standard reliability, in a standard product.

But, we do not have a policing action. If we see this, we would certainly write a reprimand type of letter to the contractor. We certainly would keep this company in mind in our future purchases, but there is no police action where we are making a constant survey of the situation.

*Mr. Pickus:* This next question has to do with becoming apprised of the various changes that take place in things like synchromotors and so forth. The question is: What is the procedure for advising the contractors of waivers to the spec that the military grants to its vendors?

*Mr. Walters:* Well, you see, this question assumes first that the military is granting waivers to the component manufacturers. On the QPL list, we are granting no waivers.

In the old days, when we were testing these without the point system, there was some human judgment involved. There was possibly a waiver of the shock requirement, as to what was damaged and what was not. There might have been a waiver on passing the endurance test.

Today, with the point system, we in effect allow a little degradation, but not much. Therefore, today, I might say, with the QPL list, we are not granting these waivers, but when we find some weakness in our old specification, we revise the old specification, and it's sent out as public distribution to the synchro manufacturers. We don't know how to reach all the contractors in the United States with any changes we make. We keep debating how to do this. It is not easy, because contractors' lists vary back and forth, and there must be a large group.

But today we do not feel we are granting these waivers that we were maybe four or five years ago.

*Mr. Pickus:* Just one more, and that is a particularly important point to contractors that have to change their own drawings every time the military changes, and that is: Why do they keep making minor changes to the synchro slash specs?

*Mr. Walters:* Well, again, you see, that is like, "Why do you beat your wife every night?"

We don't think we are making these minor changes now. We do know that we are revising our detailed specs to bring impedance values into them. If there is an error – and sometimes we have committed small errors – we, of course, clean these errors up.

Right now, with the QPL list, we have, in effect, frozen these specifications. The old specifications, which are not yet QPL, are being reviewed, and we are bringing these in line with impedance values.

Sometimes we are even raising our performance requirements to be in the state of the art. If we find one of our groups tries to make minor adjustments, we try to change it. So, really, I don't feel we are doing as much of this as the question would indicate.

*Mr. Baker:* I'm Baker, Computing Devices of Canada. I was wondering with the new lubricant you are using, what do you figure the recommended length of shelf life is (where the synchro can still meet its full specifications)?

*Mr. Walters:* We had this discussion up in a NATO overall standardization program recently. We have taken an overall figure of two years of shelf life before the unit is used as the minimum requirement, and we would certainly expect at least two years of shelf life on a lubricant, or longer.

*Mr. Baker:* What is the process of re-lifing? You send it back to the manufacturer for stripping or relubrication?

*Mr. Walters:* With our diester lubricants we haven't gotten back into the shelf life. We haven't had time, because we just started this program about six months ago. With the Naval Research Laboratory's experience over the last ten years with these types of lubricants and certainly based not just on synchros, we'd probably not expect any problem on shelf life.

*Mr. Ash:* Ash, Sperry Gyroscope. I'd like to ask these gentlemen what they would recommend in this particular case: If we have a system which calls for a defined MTBF in it, but yet it doesn't require that we utilize components with specific failure rate specifications in them because of the cost, and yet we know that the regular MIL specs are not satisfactory for meeting the system's MTBF, what would you recommend that can be done to get a little higher quality and to start, as you recommended, from the inception of the item by not insisting on failure-rate type specs, but still a higher quality than the MIL spec requires?

*Mr. Diamond:* I would do the following: Let's assume we talk about, again, the component which I would assume is the most widely understood – that's the motor – since we have it in washing machines, we have it in our driers, we look at it in our home all the time.

Let's ask ourselves a few questions. That is, "Where do the failures occur, and what are the failures we are most interested in avoiding?"

Well, the first group I think about in terms of avoiding are the catastrophic failures. That is, the open windings, short-circuited winding, the locked-up rotor, because at that point you may as well have your system acting as a weight on a fish line.

In terms of the other parameters, such as a change in torque of a few percent with age, or speed decreasing by five percent because of age, this doesn't cause a catastrophic failure in your system.

Too often, factors of safety are built in terms of overspecifying of the component. That is, the design engineer says, "Well, the standard component that I could use conceivably could be a 0.6 of an ounce unit, but I will specify it as an ounce." He's protected himself in one area, but every time you protect yourself in one area always look for the hooker: What am I giving up? Sometimes it's money, but many times it's reliability, because to get this extra 0.4 of an ounce we ask ourselves, "How does the manufacturer do it?"

If you are limited in the size of the component, he has to do it by putting in increased power or closing the air gap or doing certain things in the design which by its basic nature will affect, ultimately, reliability.

For instance, you may think that by saying, "I will specify this to be a 230-volt winding, and will test it as a 230-volt winding, but I am going to lick this problem by putting 115 volts on the winding," you are going to end up with a more reliable component, but there could be misreasoning involved. It could be a factor that in order to get 230-volt impedance levels, we have to go to wire such as Size 44 or 43, and we could have used a 39 wire and achieved a 115-volt operation.

This is just an example. It could be extra wire loaded into a 57-volt winding in a Size 8 as against a 25-volt winding in a Size 8. So you have to ask yourself, Where do I get my reliability aspect in the windings?

Well, I would like to look at my wire sizes, which are convenient to wind into the unit. For instance, we wouldn't want to put a zero gauge wire in a Size 8 unit. It's inconvenient.

If it's inconvenient to the manufacturer, you can bet your bottom dollar that it's going to give you some trouble in the field. It's really a barometer. When people have trouble producing it in the plant, when their rejection rates in-process are high, or the difficulty or costs are higher, you sort of begin to suspect that the unit itself – the field application – will not be as reliable. So you want to limit the wire sizes to wire sizes which are convenient to use in a given frame size, such as a 37 wire in a size 15 or 18; or a 39 or 40 wire in an 11 or an 8, and specify your impedance levels in such a way that there is a convenient pack and an ease of assembly and insertion. This will afford you the maximum reliability and will also result in the lowest cost.

# AN IN-HOUSE MANUAL OF STANDARD COMPONENTS – HOW IS IT PREPARED?

Joel G. Walker

## Introduction

Are you going shopping for some "off-the-shelf" items or commercially available components to fit a design you're working on? Let's walk through "radio row" and see what the various dealers have in their stockpiles and displays of components. We'll probably have to over-buy by 200 or 300 percent, then test every component when we get back, in order to be certain that we have those that will be required to do the job. Rejects??? Plenty of them, but that's the price that must be paid for economy parts which have no manufacturer or supplier listing with guaranteed maximum and minimum tolerances.

How about an alternative? The design engineer can contact one or more suppliers for the type of component he needs; wait for brochures, pamphlets, or spec sheets; select the parts he wants; try to decide whether he can rely on the supplier's representation; then place an order. A short time later another design engineer, in the same office perhaps, may have need for a similar or duplicate part and go through the same complete routine, except that he will probably choose another supplier, human nature being what it is. Both engineers will undoubtedly place future orders on the same supplier each dealt with originally, but will each keep up his own private reference library? Should a design engineer be his own component standards and procurement expert? We don't think so.

## How About a Manual?

Let's make available to the engineer a components standards manual put together by a qualified in-house standards organization which will also monitor new developments and keep the manual up to date. But since each design engineer has been doing his own shopping to his own satisfaction, why do we need a manual?

Components standards manuals ensure companywide availability of information indicating (1) components manufactured to relatively uniform quality standards; (2) interchangeability of components secured from different suppliers; (3) testing requirements to ensure compatibility; and (4) delineation of standards to ensure best cost control. A manual for your company's use would be specially keyed to your own needs.

### What Should a Manual Contain?

According to Webster, a manual is a book easily carried in the hand, a handbook. But our "manual" may take many forms: a handbook, a file of specification drawings, shelves of supplier catalogs, a file of printed specification outline cards, various microforms, or magnetic memory tape.

The manual, in whatever form it is prepared, should contain comprehensive reference information for those component types which are of primary interest to your company. The information should indicate the physical and electrical configuration and should, whenever possible, include a sketch giving reference dimensions and perhaps an appended table listing the detailed dimensions, tolerances, and values. The mechanical and/or electrical limits should be closely defined in order that the design engineer may know exactly what degrees of over-or-under design the components themselves will permit.

Testing requirements or procedures should be spelled out or the appropriate testing documentation reference given in order to ensure that all design engineers in the organization test in the same manner and to the same goal. For purchased items it is very desirable that sources of supply be indicated so that the designer can be assured of placing his request for components with a reputable, reliable, and established organization and thus, it is hoped, minimize the possibility of "shotgun" properties, while still taking advantage of cost savings through not over-designing.

In many cases the same family of components is available in three levels or grades of quality. Our manual should show these levels in order that design may be functionally acceptable and still priced within reason.

Level 1 would constitute items with wide tolerances, almost

universal availability, and lowest cost. Application would be in noncritical areas where service would probably never approach peak requirement and replacement would be simple with slight effect on operations.

Level 2 items would be made to somewhat closer tolerances, have some kind of quality assurance from the manufacturer, and consequently carry a higher price. These items would be of primary interest where close coupling with other components was essential, design required dependable output, the component or subassembly could be replaced with reasonable ease, the system would suffer only minor disruption during replacement, and the additional cost was a secondary matter.

Items in the level 3 category are considered to be high-reliability items. In this group the manufacturer does the testing and screening, and uses only the relatively few components meeting very rigid standards. The manufacturer can be required to furnish comprehensive specifications covering his hi-rel items and, in some instances, deliver test data for the specific part being supplied. This, of course, increases the cost of the item tremendously. The expense is justified if the component must function within extremely close design limits, operate under wide environmental variations, remain service free for the life of the assembly, and operate in a critical area where failure or malfunction could create problems out of proportion to the cost of the item itself.

### Organization

The table of contents or index should be carefully planned in order that a complete and accurate topical index is available and expressed in the language familiar to the employees of the company.

The importance of the index can not be overstressed. Collecting and storing information without a means of retrieval is a waste of time and effort too frequently practiced in various areas of industry and government. A carefully considered and planned index will furnish a usable and effective handle for locating the needed information at a time when it will do the most good.

The index may be based upon physical standards: length, width, thickness, weight, material, and similar information; basic electrical characteristics such as conductance, resistance, capacitance, etc, arranged in some logical sequence; part code numbers, if these are significant and are well known to all design engineers in the organization; commercial names for primarily off-the-shelf items; or any other convenient identification system. The index may be maintained in the form of a printed preface sheet, a separate small volume, a computer tape for machine reference, or, as will be explained later, in a manner which permits the manual to be self-indexed.

There are currently many forward-looking references being made to magnetic-tape-stored components data. These tapes could also be considered a form of manual, and queries originating even at a remote location could be answered almost instantaneously. A start has already been made in some organizations toward utilizing this machine capability, but again the method of listing and indexing is of primary importance in order that the material entrusted to the magnetic tape may be retrievable.

The basic considerations for a really useful index are primarily an analysis of the kinds of components needed, the capabilities and properties you must have; what tolerances you can live with; and with what equipment the component must function. Starting from here and branching to those questions and items of reference data peculiar to your oganization, an index can be created which will serve your purpose in almost any circumstance.

**Reproduction**

The method of reproduction of the manual will be determined primarily by the form in which the manual is prepared. As indicated above, present procedures permit many media to be used in preparing reference documentation. Most manuals are in the form of conventional loose-leaf books where the printed pages are removable and can be revised and replaced as necessity requires. These manuals require binders or covers of some sort and may be easily stored on a shelf or in a desk drawer where they are immediately available.

In some applications it has been found that using a printed card for each item serves equally efficiently. The cards are of some convenient size such as 4 by 6, 5 by 7, or 5 by 8, depending on the amount of information required by the designers. A file of this type requires a desk-top file box or perhaps even a larger floor-type file of many drawers, depending on the quantity of reference material. Of course, in filing the cards for a manual of this kind the reference sequence must first be determined, as mentioned under organization above.

Each card should show all the physical properties of a specific component. A further advantage here, beside that of size, is that if the component is one of a family, the variations and characteristics of the other members of the family could be listed on the same card, just as might be done in the larger printed-page manual. The card may also show a drawing of the part in question and, in the case of a family of items, indicate all the pertinent physical, mechanical, and electrical properties in tabular form. A manual consisting of cards of this kind is readily revised and does not require extensive reproduction as would be the case with a solidly bound volume.

With the growing impact of microreproduction in the documentation field, many reference manuals are prepared in 35-mm microfilm frames which are mounted in conventional electronic accounting machine (EAM) cards apertured to accept the microfilm. These cards offer a mechanical search facility in that they may have retrieval information punched into the card which the EAM interpreter prints as a filing heading. In addition, the punched information permits limited processing, sorting, and searching through the use of conventional EAM equipment.

Another type of microreproduction used to considerable extent in the field of manuals or catalogs is the 16-mm microfilm reel, mounted in a protective cartridge. With the assistance of a supplementary printed index, the reel may be searched rapidly on appropriate equipment to locate the required component or information. A single reel of film of this type can contain up to 2,500 standard 8½ by 11-inch pages. It is obvious that with family-type items this offers a possibility of placing information on 10,000 or more individual components in a single cartridge measuring about 4 inches square. For effective application, of course, a components manual of this type requires as carefully considered an index as does any other system.

Many do-it-yourselfers have been exposed to the microform repair parts catalogs in cartridges used by some large home appliance parts centers. This system has been found to work very effectively in consumer service, particularly when applied to parts which are not subject to a great deal of engineering redesign. This particular type of semiautomated manual contains pictures, dimensions, names, and part numbers. The basic requirements are that one know the exact name of the major equipment and the approximate age, and be able to recognize a picture of it on the microfilm reel. An exploded view gives the stock number beside each of the parts that goes into the general assembly.

The reel microform, as applied by industrial organizations and various government agencies, uses as a starting point a printed index having first-generation descriptors peculiar to the information contained on the reel of film.

### Master Documents

Obviously, any type of photographic reproduction must start with a clear, clean, and completely legible master document. Then, depending on the kind of manual system one has established, the original manual and subsequent maintenance sheets may be printed by photo offset on sulfite bond paper or on a good grade of card stock, or the masters may be sent to an in-house microfilming location or to a microfilm service organization for the preparation of either microfilm aperture cards or the 16mm cartridge-type of catalog.

However, one may encounter psychological resistance to the introduction of a manual in microform. If a microform is to be the medium used in your operation, then the standards organization should undertake an educational program to point up the financial and operational gains available through a properly planned microform system.

### Naming the Manual

The title of a manual really is a secondary matter. However, the name of a product frequently affects the ease with which it is "sold." One might call it Components Standards Manual, Components Card Catalog, High Reliability Parts Catalog, General Use Components, or Recommended Components Microfile. You will think of others specifically pertinent to your use.

### Distribution

Establishment of a distribution list for a new manual should be approached cautiously and at a management level. When a new book is issued, everyone wants a copy, no matter what the book covers. Therefore, the engineering, drafting, and purchasing department heads should be solicited for a specific list of the areas, groups, or individuals who must have components information available in order to properly and effectively perform their jobs. After the original distribution the list must be screened periodically for deletions and additions, and updating and correcting of addresses (room numbers and locations), thus keeping the very essential distribution maintenance information current.

Since the manual will be used as a primary reference source by many people in the plant, the copies placed about for general reference should not be filled with engineers' notations. The full-size paper manuals and, to some extent, also the card catalog type seem to lend themselves to this marking up. Engineers should be requested not to make any marginal notes or leave confusing hieroglyphics on the catalog to puzzle the next engineer. If the manual is in a microform, then consideration must be given to the strategic placement of appropriate viewing and perhaps print-back equipment so that the engineer may readily read the information contained on the microfilm in whatever form it may be and prepare a print for himself on which he can make notes if he wishes.

### Maintenance

To keep the manuals up to date, the engineering, drafting, and purchasing personnel should submit to their respective supervisors recommendations for additions, deletions, or changes in the manual with the reasons for the recommendation, and the gains they anticipate as a result. Recommendations considered appropriate by operating supervisors would then be submitted to the standards organization for review and company-wide integration. It must be remembered that it is the responsibility of the standards organization to submit additions, deletions, or changes to appropriate in-house organizations for their comment and thus hopefully avoid instituting any requirement or procedure which would prove harmful to some aspect of the company's activity.

When new pages are issued for the manual, or as pages are revised and redistributed, the holder of each manual should be instructed via a covering letter to discard the old information appearing in the manual and retain only the latest data for information and guidance. A historical file of additions and changes, and the reasons therefore, should be maintained by the standards organization responsible for the components manual.

### Final Disposition of the Manuals

Manuals covering any in-company parts catalogs should remain the property of the company and be returned to the jurisdiction of the standards organization when the individual to whom the reference data was assigned leaves the employ of the company. In the case of manuals furnished in the form of microfilm reels, aperture cards, or microfiche, special viewing and retrieval equipment is necessary and therefore data in this form is not so likely to be removed from the originating premises as a conventionally printed and bound document. However, since manuals of this nature generally contain guides and instructions for in-house operations and procedures, and may indicate the company's preference for the merchandise of certain suppliers, these data should not be generally disseminated. In addition, manuals in any form quickly go out of date. When removed from company premises and not maintained, they inaccurately reflect the company's standards and could create embarrassment in their references to outdated, superseded, or abandoned design concepts.

### What Will a Components Manual Do for You?

It will save engineering time for productive engineering. Design engineers would not be making components quality studies, or devoting effort to stay abreast of the rapidly changing components market.

It will ensure a standard method of quality reference understood company-wide, the elimination of outdated personal files, and other benefits of standardization.

It will be organized and maintained by your own standards organization and be entirely devoted to your company's needs.

# Part 6

# COMPANY STANDARDS DEPARTMENTS AND HOW THEY WORK

*There are probably few standards engineers who would not recognize the two standards departments described as the ones which made a great contribution to the success of their companies. Would any company desire a model for a standards department they may be well inspired by these two examples. In these articles will be found organizational and financial details for running a standards department.*

*R.G.*

# THE SPERRY MATERIALS AND COMPONENTS ENGINEERING ACTIVITY

Samuel B. Ash

Most companies today have some form of engineering standards program. Management has generally learned to recognize its value, but in many cases it has failed to take advantage of the economic benefits possible from such a program. To reap the full benefits, engineering standards must be more than documentation in terms of specifications, standards, and manuals. They must be integrated into a system that includes every phase of materials and components engineering as well. A comprehensive, well-organized, and dynamic engineering standards program can contribute to the effective and efficient internal operation of business.

The Sperry Gyroscope Company has for many years realized the importance of maintaining a strong centralized standards, materials, and components engineering organization. The company's products – guided missiles, inertial guidance systems, radar, fire-control systems, countermeasures, aerospace control systems, and other complex electronic and electromechanical equipment for all three services – have one thing in common: they all require low-cost and reliable materials and components to meet competition and achieve overall systems accuracy and reliability.

Years back, when products were relatively simple, the product engineer was able to handle an entire product and, as such, not only concerned himself with the overall product design and function, but also with establishing specifications and negotiating with vendors for the component parts, materials, finishes, and processes that went into the product. In short, he had to be a "jack of all trades" and possess a good working knowledge in each of these fields.

Today, with modern complex aerospace and weapons systems, it is not uncommon to find hundreds of engineers working on a single system, with each product engineer handling only a small but extremely important phase of the overall system. To keep up in such fields as radar, space propulsion, inertial guidance, laser technology, infra-red optics, and other highly scientific developments, the product engineer can no longer train himself in such allied fields as metallurgy, high polymer chemistry, solid state physics, heat transfer, kinematics, components engineering, standardization, and the like. He must now rely on supporting activities for these requirements.

The materials and components engineering activity at the Sperry Gyroscope Company is designed to provide these supporting services – in the fields of standardization, component parts, materials, finishes and processes, and design and procedural practices. By investing in a relatively small, centralized group of specialists in these fields, the company has assurance that its product needs are filled rapidly, effectively, and at lowest cost. It further has the assurance that continuity of effort, cross-fertilization between product groups, and uniformity are maintained.

The materials and components engineering activity at the Sperry Gyroscope Company is unique in many respects in its organization, functions and responsibilities, the services it renders, its requirements for a successful operation, and the various benefits and cost savings that it has achieved.

### Organization, Functions and Responsibilities

Materials and Components Engineering, MCE for short, together with the three supporting laboratories – the Electrical Measurements Laboratory, the Materials Laboratory, and the Environmental Test Laboratory – form an integrated facility reporting to a single manager in the Engineering and Manufacturing Services Division.

Materials and Components Engineering has about 100 people in the department, roughly 85 of whom are engineering and supervisory personnel. The department is divided into 4 major functional sections as shown in Figure 1. Each section is further subdivided into smaller sections and groups, and the materials, finishes, processes, electrical and mechanical components, and design and procedural practices handled by these groups are likewise shown in Figure 1.

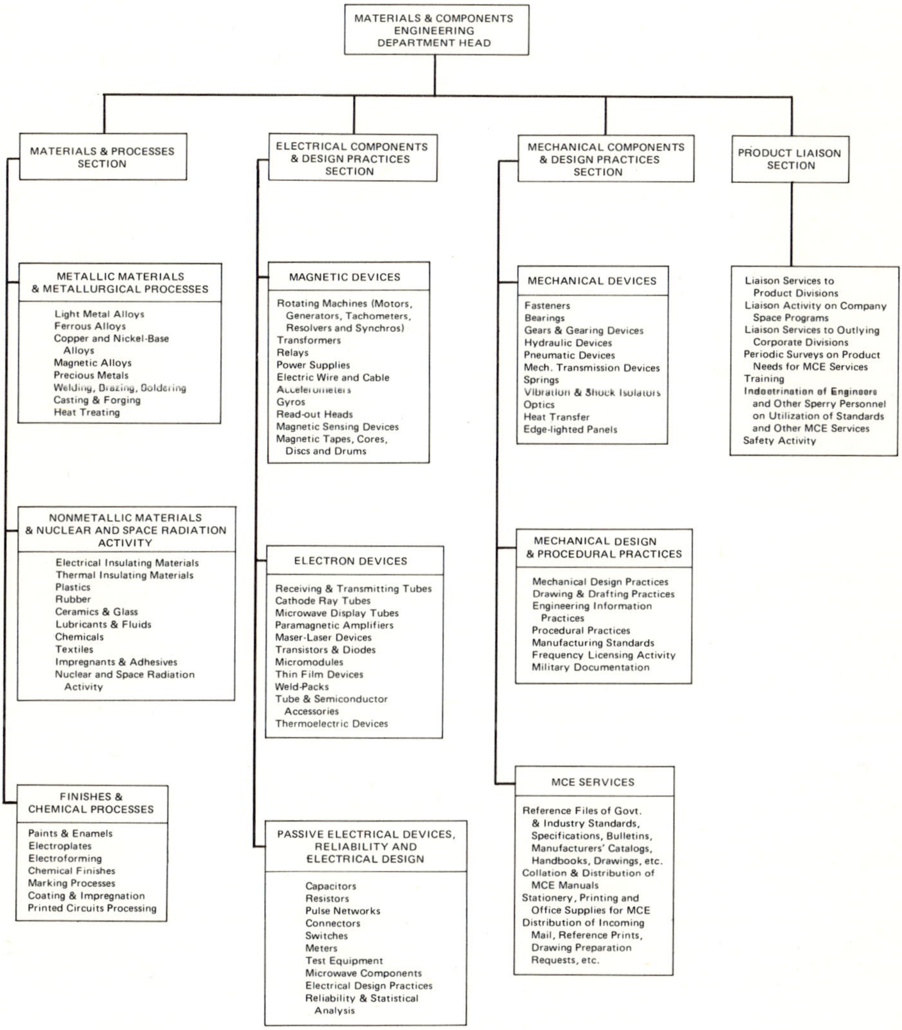


**Figure 1.** Organization of the Sperry MCE Activity.

MCE costs are mostly all indirect, with expense charged to departmental overhead. However, MCE has the capability of charging some costs direct to a product for certain functions that are specifically applicable to it – such as the preparation of purchased part drawings. The preparation of general-use drawings, standards, specifications, and other functions executed by MCE for general or multiproduct use remain chargeable to departmental overhead. About 35 to 40 percent of the MCE total work effort is expended on specific products, and about 60 to 65 percent on general work. MCE departmental overhead expense is prorated monthly to the product divisions it services. The overall departmental budget is over one million dollars per year. This represents less than 0.3 percent of the company's annual dollar sales.

MCE services five operating divisions in the New York metropolitan area: Air Armament, Surface Armament, Marine, Electronic Tube Division, and Field Engineering. There are approximately 3 MCE engineers to every 100 product engineers in these divisions. MCE also provides a limited amount of help to the outlying corporate divisions, although the latter do not share in the proration of MCE expense. While the outlying Gyro Corporate Divisions have small standards groups of their own, they utilize as many MCE-generated standards, specifications, and drawings as they can, but create for themselves anything peculiar to their own operations. MCE specialists are made available to answer inquiries on special problems on request of an outlying division.

The general functions performed by MCE are briefly as follows:

1. Standardizes component parts, materials, finishes and processes, and design and procedural practices for use in Sperry products
2. Formalizes the standards developed into written standard practice instructions (which collectively form a set of company-wide engineering standard practice manuals)
3. Develops and issues all drawings and specifications used for the purchase of materials and components
4. Develops and issues all manufacturing specifications for the application of finishes and processes
5. Develops and approves jointly with purchasing all sources of supply for materials and components
6. Furnishes technical assistance to others on the selection and application of components, materials, finishes, and processes for product use, and on the solution of engineering and manufacturing problems relating to same

7. Works closely with the technical staffs of Sperry vendors making suggestions and proposing upgraded designs to improve Sperry products

8. Plans and conducts laboratory investigations for the evaluation of new items, for qualification testing, and for the solution of engineering and manufacturing problems

9. Develops and issues quality control standards on materials and components

10. Reviews and interprets applicable government and industry standards and incorporates their requirements into Sperry practices

11. Reviews general equipment specifications at the request of product engineering and makes recommendations on compliance or deviations to be requested with regard to specified components, materials, finishes, processes, and documentation requirements

12. Participates in standardization activities of technical societies, industrial associations, and government agencies for the purpose of creating national standards and specifications suitable for use in Sperry

13. Maintains central files of government standards and specifications, vendors' catalogs, commercial standards, part number records of similar parts, performance data, test results, and samples, and related application information on components, materials, finishes, and processes

14. Keeps engineers and other company personnel abreast of the latest developments in the fields of materials, components, and design practices by conducting seminars, lectures, and demonstrations, by publishing special bulletins, and by featuring timely articles in the quarterly *MCE Bulletin.*

Combining the MCE functions into six major tasks or groups of tasks for the purpose of establishing work effort distribution, table 1 represents the averages established over a period of years.

**Table 1. Averages of Work Effort Distribution**

| *Task* | *Percent of Total Departmental Effort* |
|---|---|
| Technical assistance | 30 |
| Purchased part drawings and specifications | 25 |
| Standardization project work | 20 |
| Laboratory investigations | 10 |
| Vendor activities and development of sources of supply | 8 |
| Review of government specifications, outside committee and technical meetings and miscellaneous tasks | 7 |

These percentages may vary with the particular work load, but by and large they represent a fairly accurate distribution of departmental effort. It can thus be seen that the standardization activity itself amounts to only 20 percent of the overall effort.

### Standardization Effort

The effective standardization of component parts, materials, finishes and processes, and design and procedural practices calls for a great deal of thought and study. Consideration must be given to product needs, both present and future, customer requirements, availability, cost, internal manufacturing facilities, and other important factors. The materials and components selected for standardization must be thoroughly evaluated by laboratory tests and actual service, and must be covered by specifications for procurement and quality control.

The best standards in the world are of little value if they are not used. Standards cannot be effectively imposed by management decree. They must sell themselves by their own merits. At Sperry, the product engineer has the final say as to what goes into his product. The success of the standardization program then lies in the ability to make the engineer want to specify standard materials and components whenever possible. This can be done –

1. By making the standards of such quality and technical excellence that they will merit the respect of the engineers.
2. By placing the standardized information at engineer's fingertips.
3. By conducting a continual educational campaign.

Engineering manuals published by MCE – over 2,000 of them distributed to personnel in engineering, design, purchasing, inspection, planning, manufacturing, and stores – are the mainstay of the communications process at Sperry. The manuals are kept current through the issuance of new and revised standards on a continuing basis.

The Sperry manuals are more than compilations of standardized parts and materials by name, part number, and configuration. Technical data, reliability data, and application notes are included to permit the engineer and designer to make intelligent selection and application. Application notes are important. The highest quality product can be the least reliable if not used correctly. Experience has proved that repeat inquiries for information can be drastically reduced by publishing more data in these manuals – if not overdone. We guard against treating these manuals as textbooks.

Standardization work must be looked upon as an obligation to be met. To assure an input of at least 10 percent of the total work capacity into standardization projects, MCE at Sperry closes its doors to all users of its services every Monday morning from 8:00 A.M. until 12:00 M. Otherwise the tendency is to sacrifice standardization projects whenever demands for purchased part drawings and technical assistance are heavy. If engineering manuals get out of date they lose their effectiveness as a standardization tool.

During the Monday morning standardizing sessions no telephone messages are taken except in extreme emergencies, in which case only the supervisors can be called. Monday afternoons, perforce, call for heavy activity to dispatch business that may have accumulated as a result of the closed period. This has worked out extremely well.

MCE conducts a continuous campaign to educate engineers in the greater use of standardized materials and components. This it does by helping to indoctrinate new engineers by making presentations to various engineering groups, by publishing articles in the *MCE Bulletin* and other company publications, by issuing special memorandums to engineering, and by personal contact.

As mentioned, MCE personnel participate in outside standardization activities. MCE specialists serve on various committees of ASTM, SAE, AIA, EIA, AOA, and others. Many of the MCE specialists are recognized as authorities in their respective fields, and participate in industry-services meetings and in special panel work. MCE also participates in the Company Member Conference activities of the American Standards Association.

## Engineering Manuals

The various engineering manuals and other publications issued by MCE are briefly outlined below.

1. *Engineering Standard Practice Instructions Manual* (2 vol.). Features electrical and mechanical design practices, drawing and drafting practices, engineering procedures, manufacturing standards, military documentation requirements, definitions, abbreviations, and symbols.

2. *Standard Parts Manual* (3 vol.). Contains all of the electrical and mechanical components standardized for general use. As mentioned, in addition to the parts' numbers listing, it provides technical data, application notes, cost, reliability information, equivalent military specifications, and other data.

3. *Materials and Processes Manual* (2 vol.). Covers all of the standard materials, finishes, and processes used in the company. Includes types and forms standardized, chemical and physical properties, application notes, methods of fabrication, sizes, tolerances, applicable specifications, cost, and other pertinent information.

4. *Material Quality Control Manual.* Records the recommended practices for the inspection and quality control of Sperry incoming raw materials. It lists those incoming tests deemed necessary.

5. *Electron Tube Manual.* Contains technical data, performance, characteristics, derating curves, military specification requirements, cost, application notes, inspection and test, and other data of all the electron tubes standardized for company use. The parameters and data given are directly useable in circuit design work.

6. *Semiconductor Devices Manual.* Features information similar to that in the *Electron Tube Manual* but for transistors, diodes, solid state circuits, and other special semiconductor devices.

7. *Approved Suppliers Manual.* Lists those suppliers of electrical and mechanical components whose products have been evaluated for conformity with engineering requirements taking into consideration procurement and quality experience. Suppliers listed represent agreement between MCE and the purchasing departments, and are used as a guide in the selection of suppliers.

8. *Standard Practice Instructions for Contractors.* An abbreviated summary of Sperry standard practices, designed to aid contractors working from Sperry drawings in the procurement and application of Sperry-designated materials, finishes, and processes, and in the interpretation of workmanship requirements. It lists the approved suppliers and complete purchase information for all materials, finishes, and processes used in the company, and is therefore also employed by the Sperry purchasing departments as the official information source for the procurement of all raw materials.

9. *Non-Product Parts Manual.* Lists the electrical and mechanical components used in breadboard construction, test equipment, and for other nonproduct use. These items are stocked and maintained in a separate stockroom.

10. *Materials & Components Engineering Bulletin.* The *Bulletin*, published quarterly, features up-to-date information and announcements on components, materials, finishes, processes, and design practices not yet completely evaluated or standardized for Sperry use. It bridges the gap between information published in the engineering manuals and the very latest developments. Additionally, it brings to the readers' attention solutions to special

problems, an up-to-the-minute report of all the materials and components laboratory projects initiated or completed, and the latest government and MCE specification releases. In-between issues are published covering some special subject of general interest when required.

### Technical Assistance

When requirements are not covered by the Sperry engineering manuals, product engineers, designers, etc, are urged to utilize the services of Materials & Components Engineering. As previously outlined, assistance is given in selection and application of materials and components and for the solution of engineering, design, and production problems involving component parts, materials, finishes, and shop processes.

From the mass of information accumulated over a period of years, and from experience and specialized training in materials and components work, the MCE engineer is in an excellent position to render immediate and effective service. If he does not know the answer to a problem, he makes it his business to find it. He may make a search of the literature on the subject, consult the technical staffs of Sperry's vendors and the military establishments, and conduct laboratory investigations, among other things. The important thing is that he find the right answer. Unless the product engineers receive prompt and satisfactory solutions, they will not bother coming back again.

Experience warns that it is extremely important that product engineers contact the MCE specialists for their materials and components requirements at the earliest possible time. Unless this is done during the early developmental and breadboard stages, the product design may be "frozen." It may be too late in the schedule for MCE to verify the need for a new part or material, verify the quality, reliability, and cost, or do much about the vendor selected. If allowed to go unchecked, this practice would result in the creation of more and more nonstandard parts and materials, higher unit costs, larger inventories, more test work, and considerable duplication of engineering and design effort.

To help avoid this situation, MCE regularly reviews all new contracts received by Sperry, and arranges with product department heads for the assignment of one or more MCE engineers to be resident in the product area during the early engineering and developmental stages. The resident

MCE engineer studies the equipment requirements, reviews schematics and layouts and assists in the selection and application of as many standard materials and components as possible. He brings other MCE specialists early into the picture of the development of any needed new or improved materials and components. Early effort is needed on planning qualification and reliability tests, for reviewing documentation requirements, and for assisting in the preparation of nonstandard parts lists for customer approval. This residency in a product area and liaison activity has been of great value in cutting down the number of new parts and materials created, improving product reliability and performance, and effecting cost reduction. The salary of the resident MCE engineer is charged to the product, unburdened.

As helpful as the described technique is for the utilization of MCE services during the early phases of product development, it cannot be considered as a substitute for the high technical excellence required of the MCE engineer, in order to be of real value to product engineers. The MCE engineer must be able to analyze complex engineering problems and employ imagination and ingenuity in formulating new ideas or devising ways and means for solving these problems.

During proposals and advance studies and in early stages of product planning, the problem presented to the MCE engineer frequently is in the form of a statement, often partial, of a subsystem function to be achieved, and is usually accompanied by a proposed method conceived by the product engineer. The MCE engineer must make an analysis to determine whether the proposed method is compatible with its objectives for the types of materials and components suggested. In many cases, the MCE engineer's analysis reveals either that the original concept is unworkable with available materials and components, or that alternative approaches would offer advantages. In such cases, he creates one or more proposals, taking into account other determinant factors such as reliability, cost, and availability.

The following examples will help to illustrate this type of activity:

1. MCE was requested to review a design for an r-f drive circuit for testing highpowered klystrons. The product engineer suggested signal generation either by a klystron oscillator or by a low-level signal generator and twt amplifier combination. The MCE engineer disagreed with these approaches and demonstrated the difficulties that would result from either method. He then presented a proposal for the use of a voltage-tunable magnetron oscillator and twt combination and provided complete design data and cost information for the r-f drive circuit. The new design resulted in a decrease in cost, complexity, and maintenance problems in the equipment.

2. MCE was requested to analyze the electronic circuits for a complex navigational system in order to provide recommendations for the choice of microcircuitry techniques. The MCE engineer worked directly from 24 circuit schematics, classified each circuit as to analog, digital, or power supply function, and determined whether solid state or hybrid techniques should be selected for each circuit, with proper consideration for miniaturization, system accuracy, reliability, cost, and customer requirements.

3. Reports from the field indicated unreliable, spasmodic operation of the crown gear clutch in an aircraft navigation system. At the request of the product engineer, MCE studied this problem and it was found to be caused by "hanging-up" of the top lands of the gear teeth during engagement. By utilizing the principles of involute trigonometry, the MCE engineer was able to eliminate the top lands by designing pointed gear teeth which provide smooth, reliable, uninterrupted engagement of the clutch.

4. MCE was requested to analyze the design of a digital computer in order to come up with recommendations for its conversion to a nuclear radiation-resistant system. The MCE engineer thoroughly reviewed the existing design with the product engineers, pinpointed the materials, components, and construction details which would be adversely affected by nuclear radiation, and suggested suitable substitutions. MCE also advised on radiation exposure testing, dosimetry, and analysis of test results. The redesigned computer satisfactorily passed the radiation exposure tests.

### Purchased Part Drawings and Specifications

MCE has the responsibility for the preparation of drawings for purchased parts – specification control drawings, in military language. Since MCE is concerned with all the technical aspects of component parts – laboratory evaluation, standardization, data compilation, technical assistance, vendor relations, and quality control – MCE is in an excellent position to determine drawing requirements for procurement. Furthermore, this purchased part drawing responsibility permits MCE to carefully screen all requests for new parts and make recommendations for the use of standardized parts or other existing proven parts in their place.

Two types of purchased part drawings are prepared. One is a general-use drawing which represents the purchased component parts standardized in our engineering manuals, and other items intended for multiproduct use. The general-use drawing bears the MCE control sticker and is the sole responsibility of MCE. Only the MCE engineer approves this type of drawing and any changes to it. The other is a limited-use drawing prepared expressly for a specific product in one of the divisions. This drawing bears the control sticker of the division involved and is the joint responsibility of MCE and the product engineer for whom the drawing is prepared. Both the MCE engineer and the product engineer approve this drawing and any changes to it. Drafting facilities for the physical preparation of all purchased part drawings are supplied by a central design group in the Engineering and Manufacturing Services Division.

Whether a part is procured as a military standard part or not, Sperry always prepares its own purchased part drawing to facilitate its handling throughout the company. Each drawing gives a pictorial outline of the part, overall and mounting dimensions, tolerances, engineering requirements, procurement information, approved suppliers, the critical tests to be performed by Sperry, and other pertinent information. Every purchased part drawing has a 7-digit code number assigned to it to permit the utilization of data processing techniques in the compilation of records and associated reports.

In establishing specification requirements for the procurement of the part, military specifications are invoked whenever applicable. When Sperry has requirements over and above the MIL spec or requirements that differ from it, these additions or changes required are enumerated on the drawing. If there is no applicable military specification, next best is a national, standardized, industry specification. If none is available, Sperry issues its own specification or delineates the requirements directly on the drawing. For materials, ASTM specifications are employed extensively.

### Book-form Drawings Used

Sperry has recently adopted the book-form purchased part drawing which utilizes multiple 8½ by 11 size sheets instead of a single page sheet. Aside from a substantial reduction in cost resulting principally from the ability to type rather than hand-letter and to make changes to one sheet at a time, the main advantage of this type of drawing lies in the ability of purchasing to tear off the last sheet, containing the approved vendors and the critical tests to be performed by Sperry, before sending the drawing to a supplier. Thus the supplier no longer knows the number or identification of the vendors Sperry has for the part or the exact tests Sperry will perform. This results in more competitive prices and higher quality. The book-form drawing also facilitates handling and storing as no folding is required and permits the grouping of similar parts into books.

### Contractual and Systems Requirements Coordinated

Request for new purchased part drawings that MCE receives from product engineering via the divisional design department may reach MCE a few at a time and with insufficient information of system requirements. Modern aerospace and weapons systems are growing steadily more stringent, and unless the MCE engineer has a complete understanding of the overall contractual and system requirements, purchased part drawings prepared from individual requests may not satisfy as well as they might special environmental needs, reliability factors, customer documentation needs, and so on. To overcome this difficulty, MCE assigns a drawing coordinator, or utilizes the resident MCE engineer when there is one, to acquaint himself with the product in the early design stages, and to obtain from product engineering all the contractual and system requirements. The MCE drawing coordinator or resident engineer then composes and issues a set of instructions to all MCE engineers which enables them to produce drawings of purchased parts that are complete, uniform, and acceptable to the customer.

### Unsatisfactory Drawings Eliminating

In spite of every precaution taken to see that drawings prepared for purchased parts are complete and fully acceptable to the vendors listed, some drawings will reach purchasing which a vendor will return as unsatisfactory. Short procurement schedules do not offer the buyer sufficient time to carry on new negotiations between MCE, product engineering, and the vendor, to wait for change orders to be written, drawing changes to be made, and new prints to be supplied to the vendor.

Release of drawings that aren't entirely "clean" is caused by two factors. One is too short a time to completely verify the drawing requirements with the vendor and the other is that the vendor does not always furnish information that is accurate or that he was willing to back up when the purchase order is issued.

Two current MCE practices overcome these problems. One, when the MCE engineer is not completely satisfied with the completeness of the information at the time he is forced to release the drawing, he continues to negotiate with the vendor until complete agreement is reached. He then puts through a change to the drawing. Normally, by the time the drawing reaches the buyer, the requirements have been completely verified with the vendors. The other is that when the MCE engineer refers a drawing to a vendor for verification, a note is added to the transmittal letter cautioning the vendor that he will be held responsible for the accuracy of his information and his commitment to make the part.

### Existing Product Drawings Reviewed

Whenever existing product drawings for purchased component parts are to be utilized in a new product design, a set of these drawings is forwarded to MCE for review prior to their use. This review permits MCE to determine if an existing standard part or a newer component featuring miniaturization, greater reliability, and/or lower cost should be used instead, to update the drawing to conform to the latest customer and vendor requirements, to add additional sources of supply whenever practicable, and to value-review the part to reduce cost. Experience has shown that the small expense and time consumed in the review are outweighed manyfold by the savings and troublefree operation that result.

### Sources of Supply and Vendor Activities

Developing adequate and reliable sources of supply for materials and components is a joint responsibility of purchasing and MCE at Sperry. Purchasing is responsible for approving vendors from a procurement standpoint according to:

1. Financial structure and credit and credit rating
2. General desirability of engaging in business transactions with a particular vendor
3. Production and delivery capability
4. Price structure
5. Other economic considerations

MCE is responsible for approving vendors from a technical and customer-requirements standpoint in respect to:

6. Engineering and manufacturing know-how
7. Ability to meet specification requirements
8. Quality assurance and test capabilities
9. Customer qualifications, and others

MCE works closely with purchasing in establishing approved sources of supply. So that it is not necessary to negotiate approval from purchasing for every vendor placed on a drawing, agreement is reached ahead of time as to which vendors are satisfactory to do business with, and these are recorded for each class of components in the *Sperry Approved Suppliers Manual*. The MCE engineers then utilize these lists as a guide in developing suppliers for specific parts.

Purchasing can buy component parts only from sources listed on the purchased part drawing, and raw materials only from sources listed in the *Sperry Standard Practice Instructions for Contractors Manual*. Purchasing can, however, solicit additional sources of supply for a given item, but it must submit such sources to MCE for approval before buying from them. MCE in turn must check out the technical capabilities of the proposed sources, qualification-test representative samples, and obtain customer approval via the product engineer when contractually required.

Purchasing can divide the business among the approved suppliers in any way it sees fit. If an approved supplier complains to MCE that he has not been receiving any business, he is reminded that he is on the approved list and the rest is up to him and purchasing, and the only concern of MCE is the maintenance of the quality of the product. If the quality deteriorates, the vendor is so informed and given an opportunity to restore the product to its original quality. If the trouble persists, the vendor is removed as an approved source.

### Two or More Sources of Supply

It is the policy of MCE to list two or more sources of supply for purchased parts whenever possible. Very often, however, circumstances such as a lack of component interchangeability among vendors, high development and tooling costs, tight schedules, inability of vendors to become qualified, and other reasons make it necessary to release drawings with a single source of supply. When only one or two equipment prototypes are involved, and it is not known whether a production order will follow, it has generally been accepted that it would be economically unsound to attempt to develop additional sources at that time.

On production orders, however, which may come months or years later, the single-source drawing presents a real problem to purchasing from the standpoint of price and delivery. By the time purchasing receives the requisition for procurement, there is usually little time available to successfully develop additional sources of supply. The ferreting out of single-source drawings and the subsequent development of additional sources must be done at the earliest possible moment after the receipt of the production order. MCE helps to accomplish this by working closely with the product engineer during the early stages of the production contract, via the MCE resident engineer or drawing coordinator, and by reviewing the list of existing purchased part drawings to be utilized in the production product. Of course, the best way to solve the problem of single-source drawings is to encourage the engineer to use more standardized parts with their concomitant multiple sources of supply.

### Advantages of Buying from Approved Sources

The practice of purchasing materials and components from lists of approved suppliers has many advantages. It provides purchasing with ready sources of supply for each material or component to be procured, thereby facilitating the placing of orders and the obtaining of quick deliveries. It gives purchasing that security which comes from knowing that it is buying from vendors whose facilities and capacity to produce have been surveyed, and who have been instructed fully in regard to the requirements of the material or component. The benefits derived by MCE are likewise numerous. This practice enables MCE to exercise control over the quality of the materials and components used in Sperry products. Also, since MCE evaluates and approves all vendors, MCE comes in contact with the sales and technical personnel of the manufacturers seeking our business, and thereby receives a steady stream of information on new and improved materials and components. It also helps MCE to obtain the technical information needed for the preparation of purchased part drawings and specifications, and the samples needed for test.

Since it is economically impractical to test every material and component, MCE encourages the vendors to do as much testing on their own as possible and submit certified test results to MCE. Unfortunately, few vendors have complete test facilities, and many cannot afford to utilize commercial test laboratories. Nevertheless, some valuable test data are obtained from a number of manufacturers. Some test data on materials and components may be obtained from the military services and from technical societies and industry associations. Sperry supplements these data with substantial testing of its own.

**Laboratory Investigations**

Hundreds of new materials and components come on the market yearly that have potential application in Sperry products. However, adequate data on their reliability, performance, and usage factors are usually not available. It is therefore necessary to prove out these new materials and components by laboratory, environmental, and life tests to some degree before they can reliably be applied to product use.

Once a year, MCE prepares and budgets for a suitable yearly test program for the evaluation of new materials and components under representative environmental and physical stress levels.

The actual laboratory work is performed by personnel in the three supporting laboratories in the Engineering & Manufacturing Services Division. MCE engineers' time is applied to the planning, preparation, and release of the laboratory test outline, procurement of samples, monitoring of tests, interpretation of test results, and the writing of the final close-out report.

In addition to its own financed laboratory program, MCE plans and conducts laboratory programs for product engineering for qualification testing, reliability testing, and the solution of engineering problems – all relating to component parts, materials, finishes, and processes. The managing of materials and component test programs by MCE specialists for a specific product produces optimum results at a minimum of cost, avoids duplication of laboratory work among the various product groups, and assures dissemination of test results throughout the company.

Laboratory testing and evaluation of materials and components are time consuming and costly. It is therefore necessary to acquire as much information as possible from other sources to supplement our own test program. As mentioned, an attempt is made to utilize vendor test data when available and when deemed reliable. The various Sperry Rand corporate divisions regularly interchange test information on materials and components in order to avoid duplication of effort and cost. Sperry also makes use of components test data derived from the various government-sponsored data exchange programs such as IDEP, GMDEP, FBMWS, and others.

The MCE activity manages a limited number of materials and component test programs contracted from the outside. In the past few years, MCE has managed several large test programs for the Sergeant missile produced in our Salt Lake City, Utah, facility, and a very large test program for the Navy on miniature components. Only those programs that are allied to our own activities and can add to our basic information are considered.

### Economic Considerations and Cost Control

The MCE activity must contribute substantially to the economic posture and profitability of the business to justify its existence. At Sperry, the MCE engineer is called upon to make decisions in the daily pursuit of his functions and responsibilities which have a profound impact on the company's economic well-being.

About 70 percent of the cost of Sperry products goes into materials and components. The MCE engineer, in selecting materials and components for standardization, in guiding product engineers in the selection and application of materials and components for product use, in setting the specifications for procurement, in developing sources of supply, in formulating incoming inspection and test requirements, in improving manufacturing processes, in guarding against excess inventories, is continually making decisions affecting cost and a broad phase of the company's business.

Savings effected by MCE activities have always been very substantial, but it took the advent of the value engineering and cost reduction programs at Sperry, which documented and recorded savings, to bring MCE's contributions to light. In one year MCE engineers, working with various operating groups, produced verified tangible savings of over one million dollars. This is in addition to the intangible savings of many times that amount that accrue from a reduction of the variety of components, materials, finishes, and processes used in the operations of the company; a reduction of engineering and drafting time by initiating standards and general-use drawings; by making available information regarding the most suitable component, material, finish, or process for a specific application;

by disseminating information on the latest developments; by recording solutions to recurring problems; and by providing systematic, timely information and counsel, including interchangeability and substitution data, to aid in avoiding loss due to obsolescence and supersedure.

A few examples of the many tangible savings effected by MCE follow:

1. Substituted an oil dielectric capacitor for a reconstituted mica capacitor on a radar pulse-forming network at a net saving of $146,000
2. Developed additional sources for 16 cables that resulted in a net saving of $50,000
3. Developed new technique for lighting edge-lighted panels at a net saving of $23,000 per year
4. Convinced product engineering that it is not necessary to test for insertion loss on radio interference filters because the checks on inductance and capacitance already gave adequate information, for a saving of $15,000
5. Participated in a value engineering analysis of a radar system and made proposals which resulted in savings of $173,000
6. Established mandatory standards for hook-up wire which reduced stock by 4½ million feet valued at $258,000 resale, and produced a cost saving in maintenance amounting to $38,000 per year
7. Standardized on double-shielded ball bearings and on ABEC-5 miniature bearings for general use in Sperry with an estimated saving of $100,000 per year
8. Introduced a new, noncorrosive, activated soldering flux which cut soldering time in half, with an estimated saving of approximately $1,000,000 per year

### Summary

From the foregoing description, it can be seen that the materials, components, and standards activity at the Sperry Gyroscope Company is well organized, comprehensive, and one that cuts across every important operation of the company. It is of high technical quality, dynamic in nature, and conducive to the development and manufacture of reliable products at low cost. It helps to preserve company know-how, facilitate communications and common understanding of requirements, conserve time and effort, provide stability, uniformity, and efficiency, and contribute to the growth and profitability of the business. It offers the best means of coping with an ever-increasing complexity and multiplicity of today's technical developments.

# MANAGEMENT OF A COMPANY STANDARDS PROGRAM

Paul A. Piper

For many years, government and industry have recognized standardization as essential to aircraft, missile, and aerospace design. Much effort has been spent in the past and standardization has continued along established and traditional lines. Although these efforts may be continuing at the same rate, there has been a relative decline in effective standardization in recent years. Some excuses or causes for this apparent decline in effectiveness are:

1. Varied and often divergent requirements of the several government agencies
2. Rapid changes in technology and environmental requirements
3. Complex weapons and space systems
4. System integration with associate companies and subcontractors
5. Geographical and organizational separation of many of the aerospace contractor's operating divisions

Whereas these factors have complicated the standardization problem, they also provide the challenge and incentive for improved direction and control. The need for effective standardization is greater than ever before and the payoff of a good program can be greater in reduced costs and improved reliability.

The first factor, government requirements, is a separate subject in itself. Recent actions of the government, and particularly the Department of Defense, indicate awareness of the problem and a serious intent to arrest the decline and implement effective standardization. The other four factors are subject to substantial resolution within individual companies and through the trade and technical associations.

This paper will describe the management plan, organization, methods of operation, and control of the company-wide engineering standardization program which the Martin Company initiated in June 1961. We look upon standardization in a very broad sense. The traditional areas of drawing procedures, component parts, materials, and processes are covered. We also promote technical communication between design specialists and evaluate new developments and research as sources for provisional new practices or standards. Although our orientation is that of a government contractor with multidivision operations, it is believed that the basic objectives and many of the policies and methods should be of interest to smaller defense companies and other industries.

The Martin Company standardization program, in operation for almost two years, has resulted in substantial improvements and benefits. Many benefits have a very definite effect on company ability to perform more quickly and more reliably, although they may be difficult to measure in dollars. However, estimates have been made of the more apparent economies achieved. During 1962, estimated cost savings were \$3,524,650, a tangible return for the \$396,000 spent on the company program during that year. The 1963 standardization budget is \$585,000 and reported savings are expected to be at least 3 times those of 1962. This budget covers only the interdivisional company program. Divisional standardization activities are estimated to be 3 to 4 times that cost.

## The Management Plan

Martin always has had a substantial standardization effort. With the expansion into 6 operating divisions in 4 geographical areas, the original Baltimore standards provided a base from which the new divisions constructed their own standards. There was a strong drive for divisional autonomy, and the divisional standards increased in volume but drifted farther and farther apart. A degree of interdivisional coordination was maintained by informal and voluntary methods, but the results were variable, often slow, and ineffective. In 1960 the interdivisional standardization efforts were about one-fourth the rate of those in 1956.

Martin President, W. B. Bergen, recognized an unsatisfactory situation and issued, in December 1960, a directive calling for a company-wide standardization program.

### *Top Level Support*

The management plan was evolved by Mr. J. P. Butterfield, then General Offices Executive Engineer, now Executive Director, Technical

Operations in Orlando. As the first step, he obtained the concurrence and approval of the divisional general managers. These men appointed their chief engineers as divisional representatives for the Engineering Standardization Executive Committee.

### *Organization*

The management plan retains full divisional responsibility for the support and implementation of the Company program. The new element is the authority and direction provided by the Executive Committee and the Director of Engineering Standardization. This approach was considered more practical and efficient than the alternate of establishing a corporate standardization staff.

Figure 1 illustrates the three levels of the standardization organization. The executive committee plans and monitors the total program, the seven committees develop and administer specific programs, and the technical task groups coordinate and maintain specific standards.

### *Divisional Support*

The total standardization effort is in two categories:

*Divisional Standardization.* This activity serves the needs of the individual divisions and their direct program requirements. Costs are borne generally by divisional overhead.

*Company Standardization.* This activity is the committee and task group effort necessary to coordinate and upgrade divisional standards to company status, the preparation and maintenance of company manuals, and related interdivisional activities. This work is done in the divisions by the same personnel doing divisional standardization and is charged to divisional overhead. However, these charges are distinguished by individual account numbers so that the finance department can report what is being spent on the company program.

The basic policy is to assign interdivisional tasks to the several divisions in proportion to the size of their direct engineering effort. However, in order to maintain a balanced program it is often necessary to shift work to the smaller divisions because of the availability of special skills or because personnel in the larger divisions may be saturated with direct

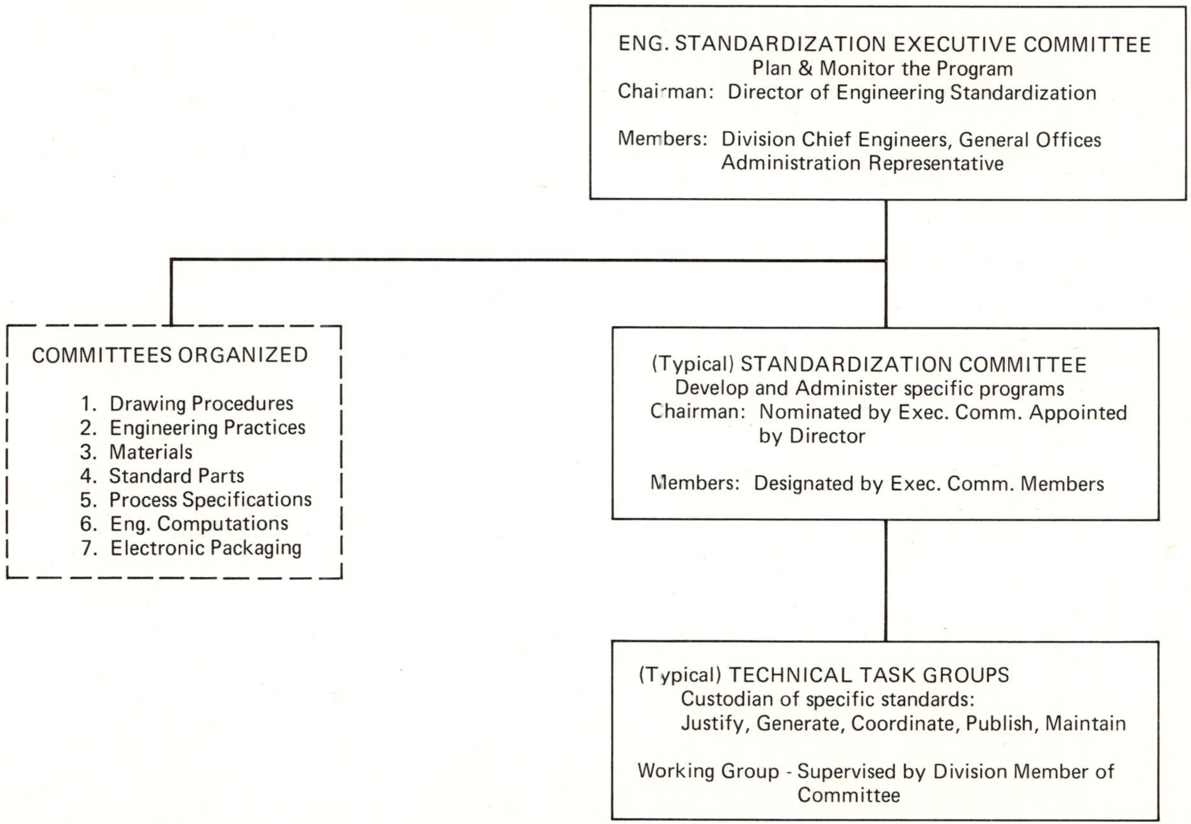


**Figure 1.** Engineering Standardization Organization.

program effort. Any imbalance of charges is redistributed each month by General Offices Finance to the divisions as additional charges or credits in proportion in their direct engineering labor base.

*Reports and Controls*

The total company organization is a matrix of 7 committees, each controlling the work of several technical task groups, with part time personnel from 6 divisions. This situation requires careful scheduling and reporting. Figure 2 illustrates the meetings and periodic reports which provide status and which generate control and corrective action as needed.

*Objectives*

Standardization is not an end in itself. Primary objectives are generally well understood. However, the company program is under continuous evaluation for its contribution to the objectives listed on figure 3. The objective of cost reduction has been mentioned above. Others need no detailed discussion except as follows.

*Advance Development.* The standardization organization does not encroach on research and development activities. However, the standardization committees provide a communication link between the divisions. For example, the Electronic Packaging Committee reviewed a new process for welding electronic leads developed in Denver. They recommended conversion to a company standard. This was done by the Process Specification Committee.

Special task groups evaluate the results of independent research activities. They issue recommended practices and provisional standards for new design requirements. Examples: *Radiation Effects Task* – Staffed by nuclear division, the group is collecting and sorting all available data on radiation environments and effects on materials and components. They provide summary reports and recommended practices to the other divisions. Also, *Thin Film Task* – Collection and evaluation of data from research reports and notebooks. The group will issue preliminary process drafts and proposed typical circuit standards for new design. These examples illustrate the policy of "standardizing ahead."

*Transfer of Work.* Multidivision operation frequently requires

| ITEM | PREPARED BY | FREQUENCY | REMARKS |
|---|---|---|---|
| Plans, Budget, Schedules | Committee Chairmen | Annual | Executive Committee Approval. |
| Committee Progress | Committee Chairmen | Monthly | Schedule status to Director. |
| Financial Progress | General Offices Finance | Monthly | Actual costs to Director. |
| Summary Program Status | Director | Monthly | Status to Executive Committee. |
| Quarterly Program Progress | Director | Quarterly | To Company and Division Mangement. |
| Committee Meetings | | 3 or 4 per year | Director attends all meetings. |
| Executive Committee | | 2 or 3 per year | Called by Director. |

**Figure 2.** Standardization Reports and Controls.

| | |
|---|---|
| 1. Cost Reduction | a. Reduce the number of products, procured parts and materials (Promote savings by volume production, procurement, inventory, maintenance and servicing).<br>b. Eliminate parallel or duplicating development in the operating divisions. |
| 2. Reduce Lead Time | Reduce the time for special design, test, and procurement by providing good standards applicable to current and anticipated needs. |
| 3. Reliability | a. Build intrinsic reliability by maximum use of the minimum number of diverse items, processes, and design practices.<br>b. Reduce errors by providing established requirements for test and qualification of parts, materials, and processes. |
| 4. Advance Development | a. Exchange development information between divisions and avoid duplicate development.<br>b. Provide stronger effort on specific developments than would be possible with independent action by each division.<br>c. Collect and evaluate research data, issue provisional standards for new design requirements. |
| 5. Transfer Between Division | Provide uniform engineering practices and documentation to facilitate interdivisional work. |
| 6. Relations with Government & Industry | a. Provide coordinated negotiations of new customer regulations and ensure proper reaction to changes accepted by the company.<br>b. Provide coordinated participation with government, industry and technical groups. |

**Figure 3.** Standardization Objectives.

design in one division and fabrication in another. Formerly the differences in drafting practices and identification of parts, materials, and processes created confusion and delays in manufacturing and procurement. Uniform engineering practices and documentation are improving this situation.

*Government and Industry Standards.* *Company Standard Parts Manuals, Materials and Process Specifications* follow the order of preference of MIL-STD-143. Government and industry standards are utilized whenever possible.

**Accomplishments**

A thorough review of the accomplishments of the 7 committees is beyond the scope of this paper. However, an explanation of the general activities, with emphasis on management aspects, is provided.

**Materials**

The work of this committee is a clear illustration of the several steps necessary for complete standardization in any area. These steps are: (a) uniform language, (b) provision for new requirements, (c) screening and eliminating obsolete or unnecessary items, (d) design usage control. Figure 4 provides dramatic evidence of the committee accomplishments.

*Uniform Language* (Uniform specification format, nomenclature and coding system for all divisions). In 1959 the 3 major divisions had 30,000 items in their separate approved materials lists. By direction of the company president (always a proponent of standardization) a central control group was established in Baltimore. This group, working with the divisional materials engineers, established a uniform system eliminating duplicating specifications and codes for identical materials. In January 1960, the total list of items available for design usage had been reduced to 17,000.

*Provision of New Requirements.* No one can accuse the materials engineers of not keeping up with new design needs. In 3½ years from January 1960 through June 1963, 14,200 new items were added, an increase of 79 percent in the gross materials code list.

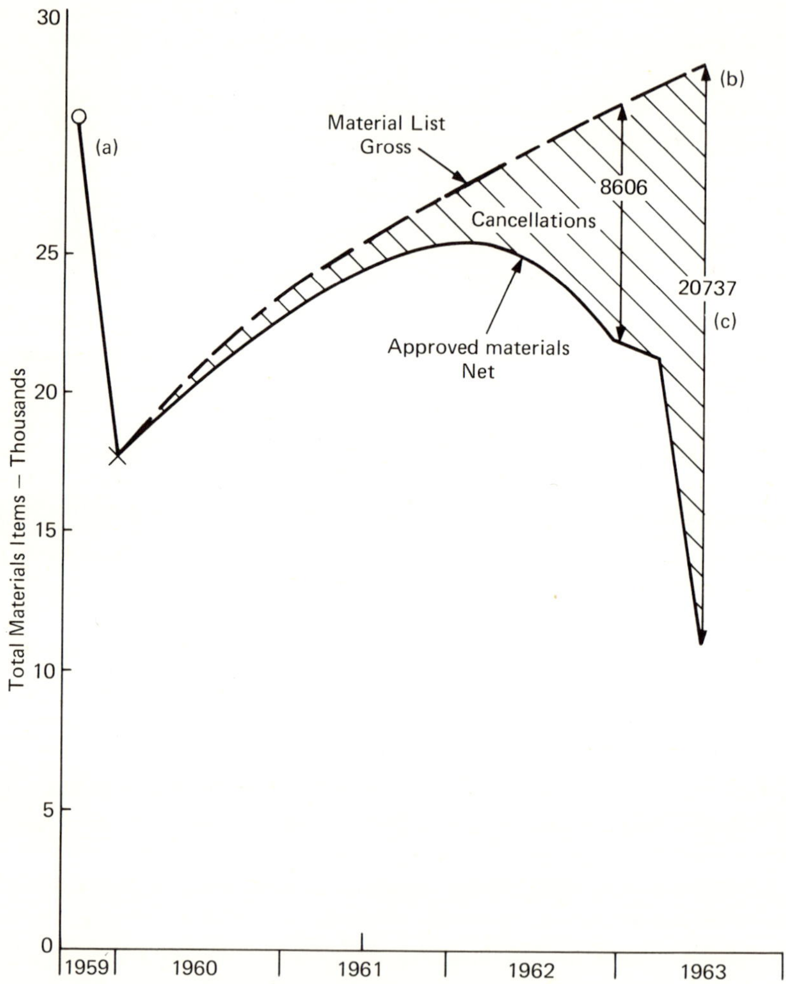


**Figure 4.** Materials Standardization.

*Screening and Eliminating.* This activity was initiated in January 1962 by direction of the executive committee. Technical task groups worked with procurement personnel to determine actual inventory usage in the divisions. Then, in consultation with designers, they proceeded to eliminate many materials specifications serving similar or identical requirements. Further, they screened the remaining approved specifications and eliminated unnecessary intermediate types, grades, and sizes. In June 1963, the total cancellations were 20,700, a reduction of 68 percent in the Gross Materials Code List. Total material codes available for new design are now 11,300 items.

The significance of this action in permitting extensive cost savings in materials procurement should be obvious. In 1962, Denver estimated procurement savings of $1,577,000.

*Design Usage Control.* The total company-approved materials list is not used on individual projects. Predesign consultations between materials engineers and designers establish a more limited, project-approved materials list. At that time the need for any new materials is established. Designers must work to the approved project list or get materials approval for new items to be added.

A new control feature has been possible with the Titan III mechanized engineering data system. A separate computer-printed bill of material is released with the picture drawing. The computer master tape file contains a coded approved materials list. If a designer tries to release a drawing with an odd material, the machine prints out a mismatch and the problem must be cleared before release of the drawings.

*Standard Parts*

The work of this committee followed a pattern similar to that of the materials committee. Some degree of interdivisional standardization had been achieved prior to the formal company program. However, the separate divisional manuals contained many duplications and detail variations.

The first major task, started in January 1962, was the selection of half the 4,400 outstanding divisional standards drawings for technical review and conversion to company standards. Figure 5 shows the actual progress through July 1, 1963. The projections to the end of the year are considered accurate. By the end of the year, 200 new divisional and corporate standards

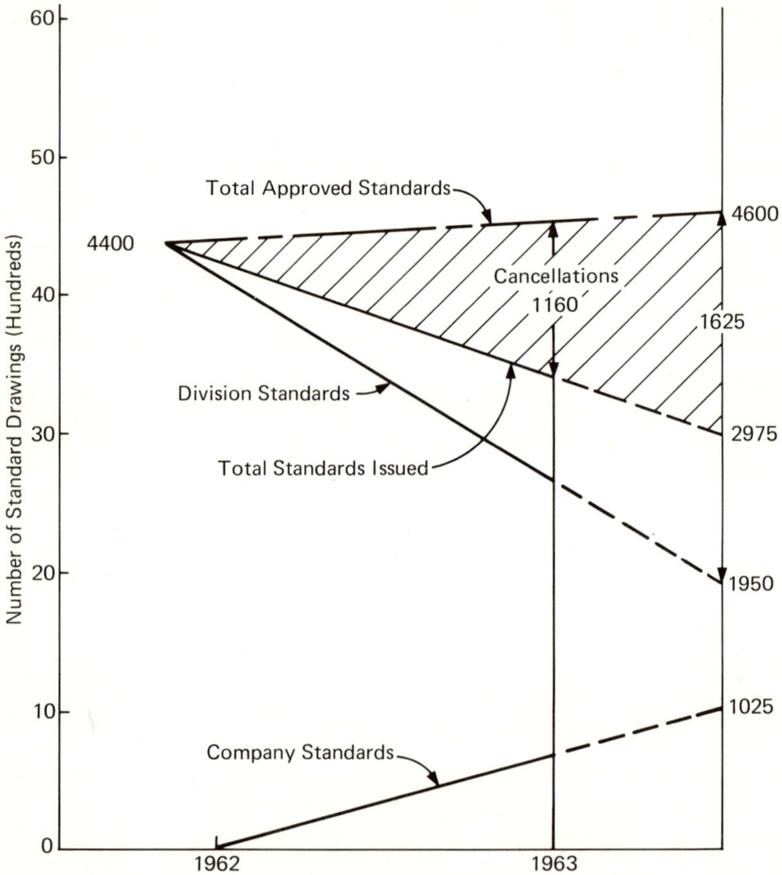


**Figure 5.** Standard Parts Drawings.

will be issued. The approved standard drawing list will contain 1,025 company standards and 1,950 divisional standards. Altogether, 1,625 divisional standard drawings will be cancelled, a reduction of 36 percent in the total drawing list. This activity will reduce the maintenance of manuals in the 3 divisions approximately $72,000 a year.

Concurrent with drawing elimination, unnecessary types and sizes were eliminated from the company standard drawings being issued. The 2,250 drawings under intensive technical review listed 154,000 individual parts. Completion of the conversion will eliminate 28,000 parts, an 18 percent reduction. This is less than the drawing reduction because many of the divisional drawings provided duplicate identification of identical parts. Continued effort next year will be directed toward further reductions in the divisional standards and elimination of actual parts.

Nonstandard purchased parts are designated as controlled commerical items (CCI). The drawings contain sufficient dimensional performance and test information to suit the intended immediate application. In the event of multiproject or interdivisional use, the CCI is easily upgraded to a divisional or company standard.

The second major task is the review of some 6,000 CCI drawings. Preliminary indications are that fewer than 5 percent of these have sufficient potential use to justify conversion to company standards. Reasons for this situation are:

1. Equivalent or superior parts now available in the present body of standards
2. Program-peculiar applications with no substantial future use expected
3. Former inadequate control of the issue of CCIs

Design usage control methods are identical to those used for materials. Predesign consultations establish a project-approved parts list.

Guide rules and checklists have been established for determining when an item should be made a standard. However, the main reliance is on the judgment of the components application engineers working with the designers.

### *Drawing Procedures*

The company drawing procedures manual was issued in June 1962. The divisions are permitted to insert special pages covering deviations or program-peculiar requirements. The new projects are discovering less need for deviations than the older ones.

The committee also has cognizance of mechanized engineering data systems. The initial system is in operation for Titan III in Denver and Baltimore. The committee is developing improvements and adaptations for new programs.

In addition to the apparent benefits of uniform basic drafting systems and each of the three major divisions publishing and maintaining one-third of a manual, other advantages derive from the direct communication and shared effort of the committee. For example: The special Denver manual on Field Site Construction Drawings was adopted and used without change when the need arose in the Orlando and nuclear divisions.

*Engineering Practices*

This committee is responsible for preparation of the *Engineering Practices Manual*. The purpose is to provide basic hardware design information for the general designer. The initial issue, July 1963, contains data on sheet metal, castings, machining, etc. Other sections on mechanical, electrical, and electronic design are in preparation.

Another responsibility is the maintenance of technical communication by means of design notes. These are informal descriptions of promising new practices or examples of poor practices to be avoided. The committee distributes these notes to interested people in all divisions. At 6-month intervals, the design notes are reviewed for possible inclusion in the *Engineering Practices Manual*.

The special task group standardizing ahead on radiation effects was described above.

*Process Specifications*

Formerly each division had separate standard processes. There was some informal exchange of process information but there were many cases of detail variations and duplicate development. Drawings from one division, released for manufacture in another, required reviews, cross-reference lists, and process deviations and departures.

This committee's responsibility is to consolidate the existing divisional processes into engineering process specifications which are applicable in the divisions and which provide better control of vendors and subcontractors.

In January 1962, there were 1,061 active divisional processes. Of these, 132 are now inactive, 222 will remain active, and 707 are to be converted. In July 1963, 101 engineering process specifications had been released, replacing 250 divisional processes. The task is about one-third complete.

*Engineering Computation Procedures*

This committee had an effective standardization program in being prior to the start of the formal company program. They have planned and achieved uniform hardware installations, and common language and programming methods in the three data centers.

A catalog of scientific abstracts is maintained in all divisions of the company. Engineers having a computation problem can locate and use similar existing programs in any division. During 1962, 75 such programs were exchanged at an estimated cost avoidance of $600,000; 77 programs were exchanged in the first 6 months of 1963 at a saving of $577,500.

The high-speed data link between Denver and Baltimore, in operation since December 1962, is an outstanding example of the value of standardization. Program inputs from Denver are worked in Baltimore and the outputs returned to Denver over leased telephone lines at night. This capability permitted a delay in renting an additional computer in Denver for a 4-month net saving of $225,000.

*Electronic Packaging*

This committee performs an integrating function with respect to the design and fabrications of electronic equipment. They have established interdivisional agreements on printed circuit-board sizes and materials, preferred module case sizes, welding and soldering processes, potting and sealing materials, and recommended design practices. These recommendations are turned over to the appropriate standardization committees for publication as revised or new standards. This is a case of designers telling the standards people what they want in the way of new standards.

The special task group standardizing ahead on thin film circuits was mentioned above.

*Cost Savings*

This subject has been cited as a primary objective. Formal and detailed cost justifications of the program are not considered essential.

However, as time permits, the committees are expected to make estimates of the more apparent economies achieved. Some examples reported in 1962 are listed below.

*Standard Parts.* Procurement of a single part (or material item) requires a purchase order, inspection, storage space, and handling time. Value engineering estimates place the unit processing or system cost at $90 to $120. The elimination of unnecessary types and sizes of high-usage parts, such as resistors, semiconductors, and mechanical hardware, resulted in systems savings of $346,650. The estimate does not include the potential saving in actual item costs due to volume procurement.

*Materials.* System savings due to reduction in the number of items plus a partial evaluation of actual material costs resulted in a 1962 saving of $1,955,000.

*Electronic Packaging.* Exchange of development data between divisions eliminated duplicate development for an estimated 1962 saving of $260,000.

*Engineering Computation Systems.* Exchange of existing programs and the data link (discussed above) saved, in 1962, $885,000.

Total estimated 1962 savings is $3,524,650. The cumulative and repetitive benefits should provide cost savings about 3 times this figure in 1963.

**Problems**

Implementation of an extensive and dispersed standardization program is not without some delays and frustrations. Anyone initiating and managing a program of this nature must be alert for possible roadblocks in his organization. In retrospect, we recognize two basic problems which might have been alleviated by better planning, wider discussion, and earlier use of the decision mechanism.

*Division Support*

Although the 1962 program was well defined by the committees and approved by the executive committee in December 1962, manning and milestone schedules showed serious slippage. During the first quarter of 1962, actual expenditures and schedule performance were 45 percent of plan. The divisional yearly plans, overhead, and manpower limitations had been established previously without consideration of company standardization. Intermediate supervision, with the best of intentions, had to give priority to identified research tasks and direct program support. Company standardization got what was left. These facts were presented to the president and general managers in April and specific instructions were issued to get back on the plan.

Recovery was not immediate. Continuing discussion of program objectives and requirements with department and section managers and administrators was necessary. Performance rose to 60 percent in the third quarter and 98 percent in the fourth. The 1962 totals were: expenditures, 71 percent; schedule performance, 76 percent of plan.

The 6-months 1963 performance was somewhat better, but the problem requires continuing attention. Expenditures and schedule performance were 73 percent of plan. A recovery plan issued in July, with transfers of work and budgets between divisions, should assure reaching established goals by the end of the year.

The solution for 1964 is to get company standardization out of the miscellaneous, and clearly identified in the divisional yearly plans. Divisional planners were given a proposed budget distribution in August, although detailed plans and schedules will not be completed until November.

*Developing a Company-wide Viewpoint*

This has not been a severe problem in most of the committees. After a couple of meetings and some remarks by the director, the members settled down to effective team work.

The exception, with disappointing progress, is the Process Specification Committee. This was puzzling because the same people, acting as the Materials Committee, have done such an outstanding job. The apparent reason is that materials coordination is entirely within engineering, whereas it is our policy that process specifications must be a joint development with manufacturing and quality. This requires three times as much coordination in each division. For a year there were extended arguments about the proper amount of detail procedures and controls in a process specification. Many specifications went through three or four coordinations and rewrite cycles before complete sign-off. This is not to imply that manufacturing or quality personnel are more prone to local pride than engineers. The problem was caused by insufficient early discussion and explanation of objectives with the other departments. This oversight had to be corrected by continuing discussions with the people responsible for review and approval.

Recently a joint meeting with engineering, manufacturing, and quality representatives from three divisions settled differences and obtained agreement on five specifications which had been in the system for several months. This practice is not recommended for normal jobs, but two or three such meetings a year earlier might have reduced misunderstandings and delays in the normal operation. Another source of improvement is the establishment this year of interdivisional manufacturing and quality committees. These groups have primary concern with their own functions but they provide a recognized means of communication, agreement, and decision making on matters of mutual interest.

### Conclusions

Management of a company standards program must generate support, understanding, and acceptance at all levels of the company. This is not difficult at the top management level or at the work level. The intermediate groups of supervision, administration, and design may be preoccupied with other things. They must understand and be involved in the effort.

A simple organization with definite delegation of responsibility, positive lines of communication, and means for establishing policy and resolving problems is essential. With a dispersed standardization effort, planning, scheduling, reporting, and follow-up are necessary to assure progress and continuity in the program. The program must not be allowed to degenerate into "documentation of the obsolete." Positive attention must be given to standardizing ahead to meet the needs of new products and programs.

During two years of the formal Martin program, substantial progress has been made in meeting basic objectives. There remains much to be done, but the standardization decline has been turned into a very healthy revival.